New Splitting Iterative Methods for Solving Multidimensional Neutron Transport Equations

Jacques Tagoudjeu

DISSERTATION.COM

Boca Raton

New Splitting Iterative Methods for Solving Multidimensional Neutron Transport Equations

Dissertation.com
Boca Raton, Florida
USA • 2011

ISBN-10: 1-59942-396-0
ISBN-13: 978-1-59942-396-8

UNIVERSITY OF YAOUNDE I   UNIVERSITE DE YAOUNDE I
FACULTY OF SCIENCE   FACULTE DES SCIENCES

DEPARTMENT OF MATHEMATICS
DEPARTEMENT DE MATHEMATIQUES

NEW SPLITTING ITERATIVE METHODS FOR SOLVING MULTIDIMENSIONAL NEUTRON TRANSPORT EQUATIONS

THESIS

Submitted and defended publicly in fulfilment of the requirements for the award of the
Degree of Doctorat/Ph.D in Mathematics

Option: Numerical Analysis

by

Jacques TAGOUDJEU

D.E.A in Mathematics

In front of the Jury

PRESIDENT :

- **Pr. Gabriel NGUETSENG**, University of Yaoundé I, Cameroon,

ADVISOR :

- **Pr. AWONO ONANA**, University of Yaoundé I, Cameroon,

MEMBERS :

- **Pr. David BEKOLLE**, University of Yaoundé I, Cameroon,

- **Pr. Nicolas ANDJIGA**, University of Yaoundé I, Cameroon,

- **Pr. François TSOBNANG**, ISMANS, France,

- **Pr Abdelghani BELLOUQUID**, University Cadi Ayyad, Morocco.

Year 2010

Dedication

*This work is dedicated to the late memory of my father **TAMBANDUE Augustin**, and to my mother **NGONYIM Sarah**.*

Acknowledgement

Here I would like to express my sincere compliments to all who have contributed from far and close to the realization of this work.

I would like to express my sincere thanks to Professor Awono Onana, my supervisor, for his many suggestions during this research. I owe to him an enormous debt of gratitude for his tireless support throughout these last years. His guidance and broad perspective has always guided me to a right path. I am very fortunate, very proud, and very honored to be Awono's student.

I am thankful to Professor Gabriel Nguetseng who expressed his interest in my work in its early stages. He has always encouraged me and has accepted to review my work and chair the thesis committee. His careful reading of the manuscript and insightful comments and suggestions help me improve significantly this work. I am also thankful to Professor Abdelghani Bellouquid who has accepted to review my work and participate to my thesis committee. His helpful suggestions have improved this work enormously.

I am grateful to Professor David Bekollé, Professor Nicolas Andjiga and Professor François Tsopnang who have accepted to be members of my committee despite their extremely busy schedules.

I would like to acknowledge Professor Norman J. McCormick and Professor Edward W. Larsen for graciously providing me with some of their reprints which have been of great help in writing this thesis.

This research was carried out, at Ecole National Supérieure Polytechnique of the University of Yaoundé I. I wish to thank friends, school faculty and staff for making my time there a great experience. I also thank the faculty and staff of the Department of Mathematics of the Faculty of Science.

I wish to acknowledge partial support of this research by the African Millennium Mathematical Science Initiative (AMMSI) through a postgraduate scholarship award. I express my gratitude to Professor Bitjong Ndombol who made it possible.

My deepest gratitude goes to all my family for their love, constant support and encouragement during my studies. Special thanks go to my brothers Tchoupou Jean-Claude and Egoko Augustin for all what they have done for me. I extend my thanks to my friends from here and abroad who helped me in hard times and shared the good times with me.

Finally, I thank my wife Jane Yuego for her love, patience and encouragements. She has been very supportive through the difficult and enjoyable times during the last years of this thesis.

Abstract

This thesis focuses on iterative methods for the treatment of the steady state neutron transport equation in slab geometry, bounded convex domain of $\mathbb{R}^n$ ($n = 2, 3$) and in 1-D spherical geometry. We introduce a generic Alternate Direction Implicit (ADI)-like iterative method based on positive definite and m-accretive splitting (PAS) for linear operator equations with operators admitting such splitting. This method converges unconditionally and its SOR acceleration yields convergence results similar to those obtained in presence of finite dimensional systems with matrices possessing the *Young property A*. The proposed methods are illustrated by a numerical example in which an integro-differential problem of transport theory is considered. In the particular case where the positive definite part of the linear equation operator is self-adjoint, an upper bound for the contraction factor of the iterative method, which depends solely on the spectrum of the self-adjoint part is derived. As such, this method has been successfully applied to the neutron transport equation in slab and 2-D cartesian geometry and in 1-D spherical geometry. The self-adjoint and m-accretive splitting leads to a fixed point problem where the operator is a 2 by 2 matrix of operators. An infinite dimensional adaptation of minimal residual and preconditioned minimal residual algorithms using Gauss-Seidel, symmetric Gauss-Seidel and polynomial preconditioning are then applied to solve the matrix operator equation. Theoretical analysis shows that the methods converge unconditionally and upper bounds of the rate of residual decreasing which depend solely on the spectrum of the self-adjoint part of the operator are derived. The convergence of theses solvers is illustrated numerically on a sample neutron transport problem in 2-D geometry. Various test cases, including pure scattering and optically thick domains are considered.

Keywords: Neutron Transport, Iterative Methods, ADI, Self-Adjoint, $m-$Accretive, Operator Splitting, SOR, Minres, Preconditioning, Numerical Results.

Résumé

Cette thèse est consacrée aux méthodes itératives pour la résolution des équations monocinétiques du transport des neutrons. Nous introduisons une méthode itérative générique de type ADI (*Alternate Direction Implicit*) basée sur une décomposition définie positive et $m-$accretive, pour la résolution des équations linéaires dont les opérateurs admettent une telle décomposition. L'analyse théorique de cette méthode montre qu'elle converge inconditionnellement vers la solution de l'équation considérée et l'accélération de cette methode par la méthode des rélaxations successives donne des résultats similaires à ceux des systèmes d'équations linéaires dont les matrices possèdent la *propriété A de Young*. La méthode proposée est illustrée par des exemples numériques dans lesquels on considère une équation intégro-différentielle de la théorie du transport. Dans le cas particulier où la partie définie positive de l'opérateur de l'équation linéaire est de plus auto-adjointe, une majoration du facteur de contraction de la méthode itérative est obtenue. Une analyse de la version incomplète de la méthode est présentée. Ainsi, la méthode a été appliquée aux problèmes du transport des neutrons en géométrie plane, en dimension deux d'espace et en géométrie sphérique 1-D. Les résultats obtenus montrent l'efficacité de la méthode pour ces problèmes. La méthode de décomposition auto-adjointe et $m-$accretive mentionnée ci-dessus conduit à un problème équivalent du point fixe où l'opérateur est une matrice 2x2 d'opérateurs. La méthode du résidu minimal ainsi que ses versions préconditionnées par des préconditionneurs de type Gauss-Seidel et polynomial sont appliquées pour la résolution de ce problème. L'analyse théorique montre la convergence de ces méthodes. Une majoration du taux de décroissance du résidu dépendant uniquement du spectre de la partie auto-adjointe de l'opérateur du départ est obtenue pour chacune de ces méthodes. La convergence de ces méthodes est numériquement illustrée sur des exemples dans plusieurs types de domaines en dimension deux d'espace.

Mots Clés: Equation du Transport, Méthodes Itératives, Méthodes ADI , auto-adjoint, $m-$Accretive, Décomposition d'Opérateur, SOR, Minres, Preconditionnement, Results Numériques.

Contents

List of Tables

List of Figures

List of Publications

Parts of this thesis have been published as:

1. **Awono Onana** and **J. Tagoudjeu**: "A Preconditioned Minimal Residual Solver for a Class of Linear Operator Equations", *Computational Methods in Applied Mathematics, Vol. 10, No. 2, pp. 119–136 (2010)*

2. **Awono Onana** and **J. Tagoudjeu**: "A Minimal Residual Iterative Solver for Neutron Transport Equation", *Int. J. Contemp. Math. Sci., Vol. 4, No. 34, pp. 1671-1684 (2009).*

3. **Awono Onana** and **J. Tagoudjeu**: "A Splitting Iterative Method for Solving the Neutron Transport equation", *Mathematical Modelling and Analysis, Vol. 14, No. 3, pp. 271-289 (2009)*

4. **Awono Onana** and **J. Tagoudjeu**: "Iterative Methods for a Class of Linear Operator Equations", *Int. J. Contemp. Math. Sci., Vol. 4, No. 12, pp. 549-564 (2009).*

5. **Awono Onana** and **J. Tagoudjeu**: "A SOR acceleration of Self-Adjoint and m-Accretive Splitting Iterative Solver for 2-D Neutron Transport Equation", *Math. Model. Nat. Phenom., Vol. 5, No. 7, pp. 60-66 (2010)*

6. **Awono Onana** and **J. Tagoudjeu**: " A self-adjoint m-accretive splitting iterative method for the solution of neutron transport equation in 1-D spherical geometry", *Proceeding of 9th African Conference on Research in Computer Science and Applied Mathematics - CARI-09 (Eds. E. Badouel, A. Sbihi, M.K. Assogba), pp 331-338, (2008);*

7. **Awono Onana** and **J. Tagoudjeu**: "A SOR acceleration of Self-Adjoint and m-Accretive Splitting Iterative Solver for 2-D Neutron Transport Equation", *Proceeding of the 9th International Conference JANO'9, Mohammedia-Morocco, (2008), pp 318-321.*

General Introduction

Overview and Motivation

Transport equations are mathematical models describing the transport of particles, energy, momentum or any transportable quantity. Initially Established by Ludwig Boltzmann more than a century ago in a study in connection with the kinetic theory of gas, the transport equations are ubiquitous in physics and in engineering. They model various phenomena in various domains such as nuclear reactor design, radiation transfer, meteorology, biology, radiotherapy, tomography, particle transport, stellar and planetary radiation, epidemiology, vehicular traffic and many others. The Boltzmann transport equations are nonlinear integro-differential equations that describe the behavior of the statistical distribution of particles in a given media, using seven independent variables: 3 in space, 2 in angle, 1 in energy or frequency and 1 in time. In the case of neutral particles, these equations become linear. The neutron transport equation is a particular generic case. In many situations, the solution of the neutron transport equation is a rapidly changing function of the spatial, angular and energy variables. Moreover, the neutron transport equation behaves like totaly different equation types from a physical situation to another. It behaves like a hyperbolic wave equation in void-like regions, in scattering dominant optically thick regions it behaves like elliptic equation for steady state case and parabolic equation for time-dependent case.

Due to their multidimensionality, the asymmetry of their operators, the lack of smoothness of their solutions and many other singularities, transport equations are very arduous to solve. Therefore, their mathematical and numerical complexities continue to be subject of intense scientific attention. In this connection, there is need of original ideas and methods on the techniques to be implemented for the solution of the transport equations. Considerable efforts have been devoted this last years on the computational techniques for the numerical approximation of the solution of the transport equations.

1

There are in general two classes of computational methods for the numerical treatment of the neutron transport equation (see [47] and the references therein): the stochastic Monte Carlo methods and the deterministic methods. Additionally, hybrid methods coupling stochastic and deterministic approaches can be applied. Each approach has its strengths and weaknesses.

The principle underlying the Monte Carlo methods is to ignore the mathematical description of the transport problem, and to directly simulate a large number of particle histories [47]. The Monte Carlo methods use the basic concept of following the paths of particle packages by randomly selecting the new directions and energies of the particles, to estimate average behavior of particle in phase-space. They work in principle for any problem. They are suitable for solving transport equations in media where the mean free paths are not very small [74]. There are no discretization errors in Monte carlo simulation, but statistical errors occur, due to the fact that the simulation ends after N histories have been processed [47]. Unfortunately this method is very time-consuming, especially in three dimensions, because the error of the results decreases very slowly [69].

In deterministic approaches, the principle is to ignore the random aspects of individual particle histories and solve the transport problems, by converting these problems into large system of algebraic equations and solving the resulting system of algebraic equations [47]. The first step of deterministic methods consists in discretizing the transport equation with respect to each of its variables : The energy variable is often discretized by multigroup approximation, in which the energy range is divided into energy groups and the interaction cross sections are approximated by histograms in energy, each histogram having one value within each energy group [49, 34]. This yields the multigroup time-dependent or time-independent according to the equation initially considered. The treatment of the time-dependent transport equation can be done by using varieties of methods such as finite difference and variational methods [2]. Finite-differencing the derivative in time is the widely used approach. The energy and time discretization lead to a coupled system of steady state single energy equations which only depend on the spatial and the angular variables. The angular discretization is often accomplished by the integral method, discrete ordinates method (S_N) and the expansion of the angular flux in terms of angular basis functions such as spherical harmonics method (P_N), Walsh functions, wavelet functions and others (see [49, 34, 27, 26, 28, 78] and the references therein). The finite difference methods, finite element methods, nodal methods and the method of characteristics are

2

usually used for the spatial discretization (see [2, 3, 4, 18, 19, 49, 48, 29, 34, 23, 61, 60]). The second step in deterministic approaches consists in solving the algebraic system of equations resulting from the discretization of the transport problem. Due to the number of independent variables of the transport problem, this system is very large and thus difficult to invert directly. The solution strategy for solving the resulting system of equations have then focussed on iterative methods [47].

Literature Review on Iterative Methods

Iterative methods are widely used for solving linear operator equations (see [1, 10, 72, 43, 45, 68, 39, 54, 65] and the reference therein). The GMRES algorithm for linear equations with bounded operators in separable Hilbert space has been study in [39]. It was shown that the results of the finite dimensional case can be generalized in the continuous case if the operator is algebraic [39]. Recently, some new iterative methods for solving linear operator equations with bounded [54] and unbounded [68] operators have been introduced and analyzed. These methods make use of the adjoint operator in the transformation of the initial equation. For the particular case of neutron transport equation, there is extensive use of iterative methods for the continuous and the discrete problems (see [5, 6, 7, 34, 30, 56, 58, 66, 67, 77, 79, 85, 84] and the references therein). The standard method is the source iteration method based on a decoupling between the differential and integral parts of the transport operator. This method becomes extremely slow in the critical case (optically thick and scattering dominant regions). Several acceleration techniques of the convergence of the source iteration method such as Diffusion Synthetic Acceleration (DSA) [5, 84], Transport Synthetic Acceleration (TSA) [5], Coarse Mesh Rebalance (CMR) [88], Quasi-Diffusion acceleration [87] and multigrid algorithms have been introduced and studied [30, 56, 5, 46, 63]. Alternative methods to the source iteration approach are the Krylov subspace iteration methods such as Conjugate Gradirnt (CG), Generalized Minimal Residual (GMRES), Bi-conjugate Gradient Stabilized (BiCGSTAB) and their preconditioned versions [5, 47, 66]. The Distribution Iteration (DI) methods based on reducing the global transport equation into coupled-cell partial current that can be solved directly [36], have been applied. The angle space distribution iteration method which combines a non-linear, high angular-resolution flux approximation within individual spatial cells with a coarse angular-resolution flux approximation that couples all cells in a

3

spatial mesh, has proved its efficiency for slab geometry problems [83]. In [62], a new iterative method based on the idea of dividing the transport solution into its particular and homogeneous components was successfully implemented in slab geometry with isotropic scattering and one energy group. Based on the natural splitting of the integral part of transport operator, other methods such as Jacobi, Gauss-Seidel [79] and Successive over-relaxation (SOR) iteration have been successfully applied to transport problem by solving a fixed point problem derived from the source iteration method. Using the same splitting, an adaptation to the infinite dimensional case of the minimal residual iteration method (see [6, 7]) has been proposed for the solution of the transport in slab geometry, in 2-D cartesian geometry and in 1-D spherical geometry. This method has been proved to be efficient and it competes with the SOR method. Further, its preconditioned versions have been analyzed (see [80]).

Thesis Objectives and Results

In this thesis, focus is given on iterative methods for the numerical treatment of the single group steady state neutron transport equation in slab geometry, bounded convex domain of $\mathbb{R}^n$ ($n = 2, 3$) and in 1-D spherical geometry.

We introduce a generic ADI-like iterative method (see [59]) based on positive definite and m-accretive splitting (PAS) for linear operator equations with operators admitting such splitting. As mentioned above, theoretical results show the convergence of the method and its SOR acceleration yields convergence results similar to those obtained in presence of finite dimensional systems with matrices possessing the *Young property A* (see [50, 89]). The proposed methods are illustrated by a numerical example in which an integro-differential problem of transport theory is considered. In the particular case where the positive definite part of the linear equation operator is self-adjoint, an upper bound for the contraction factor of the iterative method which depends solely on the spectrum of the self-adjoint part is derived . As such, this method has been successfully applied to the neutron transport equation in slab and 2-D cartesian geometry and in 1-D spherical geometry.

The self-adjoint and m-accretive splitting leads to a fixed point problem where the operator is a 2 by 2 matrix of operators. An infinite dimensional adaptation of minimal residual and preconditioned minimal residual algorithms using Gauss-Seidel, symmetric

Gauss-Seidel and polynomial preconditioning is then applied to solve the matrix operator equation. Theoretically, the methods are shown to be unconditionally convergent and upper bounds of the rate of residual decreasing which depend solely on the spectrum of the self-adjoint part of the operator are derived. The convergence of theses solvers is numerically illustrated on a sample neutron transport problem in 2-D geometry. Various test cases, including pure scattering and optically thick domains are considered.

Thesis Organization

The remaining of this thesis is structured as follows.

In Chapter 1, the presentation and the properties of the neutron transport equation are given. After a physical derivation of the neutron transport equation, existence and uniqueness results are presented for the time-dependent and the time-independent equations. Additionally, some alternative forms of this equation such as integral form, second order forms and the diffusion approximation, which can reduce the complexity of the initial first order integro-differential transport equations in certain circumstances are presented.

In Chapter 2, usual ways for the discretization of the transport equation in energy, time, space and direction are are briefly discussed. Some iterative approaches for solving the transport equations are presented.

In Chapter 3, focus is given on iterative methods for the numerical treatment of the single group steady state neutron transport equation in slab geometry, bounded convex domain of $\mathbb{R}^n$ ($n = 2, 3$) and in 1-D spherical geometry. These methods are ADI-like iterative method based on positive definite and m-accretive splitting (PAS) for linear operator equations with operators admitting such splitting. we present the PAS iterative method. The convergence analysis of the method and its SOR acceleration is provided. The convergence of the proposed methods are illustrated by a numerical example in which an integro-differential problem of transport theory is considered. Next, the convergence analysis of the Self-Adjoint and m-Accretive (SAS) iterative method and its incomplete version are presented and the convergence of the method is numerically illustrated and compared with the standard Source Iteration method and multigrid method on sample problems in slab geometry and in two dimensional space. Further, we introduce and analyze an infinite dimensional adaptation of a minimal residual algorithm linked to the self-adjoint and m-Accretive Splitting. Comparative numerical results are presented for

5

a sample neutron transport problem in 2-D geometry. Finally the convergence of the previous minimal residual algorithm with Symmetric Gauss-Seidel and polynomial preconditioning is established and comparative numerical results are presented.

Chapter 1

The Neutron Transport Equation

This chapter is devoted to the presentation and the properties of the neutron transport equation. After a physical derivation of the neutron transport equation, existence and uniqueness results are presented for the time-dependent and the time-independent equations. Additionally, some alternative forms of this equation such as integral form, second order forms and the diffusion approximation, which can reduce the complexity of the initial first order integro-differential transport equations in certain circumstances are presented.

1.1 Introduction

The evolution of a system of particles in a given domain is characterized by the particle distribution function f, which is a positive function depending on time t, the position $\mathbf{x} \in \mathbb{R}^d$ and the velocity $\mathbf{v} \in \mathbb{R}^d$ of particles. This distribution function describes the statistical evolution of the system of particles. It must satisfy :

$$f(t, \cdot, \cdot) \in L^1_{loc}(\mathbb{R}^d \times \mathbb{R}^d), \qquad (1)$$

where $L^1_{loc}(\mathbb{R}^d \times \mathbb{R}^d)$ denotes the space of functions $u : \mathbb{R}^d \times \mathbb{R}^d \to \mathbb{R}$ such that the restriction of u to any compact subset of $(\mathbb{R}^d \times \mathbb{R}^d)$ is an integrable function [25]. Therefore, the quantity $f(t, \mathbf{x}, \mathbf{v})d\mathbf{x}d\mathbf{v}$ represents the probability of finding particles in an element of volume $d\mathbf{x}d\mathbf{v}$, around the point $(\mathbf{x}, \mathbf{v})$ at time t.

The distribution function f is governed by a particle transport equation or kinetic equation. For each system of particles there exists particular types of kinetic equations. The specific form of these equations is determined by the nature of the considered system

7

(gas, solid, liquid, etc...), the nature of the interactions between the particles and the values of parameters fixing the macroscopic state of the system (density, temperature). In many cases, these equations can be written in the form

$$\frac{\partial f}{\partial t} + \mathbf{v} \cdot \nabla_x f + F(t, \mathbf{x}) \cdot \nabla_v f = \left(\frac{\partial f}{\partial t}\right)_{coll} + s, \tag{2}$$

where ∇_a ($a = x$; $a = v$) denotes the gradient with respect to the variable a and $F(t, \mathbf{x})$ represents a force field to which particles are subjected. The left hand side of equation (2) represents the variations of the distribution function f due to the movement of particles; the first term of the right-hand side of (2) represents the variations of f due to the interactions of the particles between them and/or with the medium which they occupy; the second term of the right-hand side represents the particle production sources in the medium.

It is supposed that the action of the field of force is negligible. When the interactions between the particles are regarded as collisions appearing at time t, at the position $\mathbf{x}$ and producing changes only on velocity, and that these collisions are binary (utilizing only two particles), one obtains the following form of the transport equation called Boltzmann equation :

$$\frac{\partial f}{\partial t} + \mathbf{v} \cdot \nabla_x f = Q(f, f). \tag{3}$$

The operator Q is an integral operator called Boltzmann collision operator. It is generally nonlinear. When the collisions between the particles are negligible (it is the case of the neutrons, photons...), the operator Q becomes linear. In this work, we consider neutral particles.

The remainder of this chapter is organized as follows. The section 1.2 is devoted to the physical derivation of the neutron transport equation. Existence and uniqueness results are presented in section 1.3. In section 1.4 alternative forms of the transport equation are presented. We give in the following a physical interpretation of neutral particle transport.

1.2   The Neutron Transport

In this section, we give a physical description of the transport equations. Let us denote by :

- $D \subset \mathbb{R}^3$ an open bounded domain (occupied by the particles) with piecewise smooth boundary ∂D;

- S^2, the unit sphere of $\mathbb{R}^3$;

- $I \subset \mathbb{R}_+^*$, the energy interval;

- $[0, \tau_1] \subset \mathbb{R}_+$, the time interval.

The particles are assumed to travel in the seventh dimensional phase-space $D \times S^2 \times I \times [0, \tau_1]$.

At time $t \geq 0$ each particle is locatable in $D \times S^2 \times I$ by the standard phase-space coordinate

$$\mathbf{P} = (\mathbf{x}, \mathbf{\Omega}, E), \tag{4}$$

where

1. $\mathbf{x} = (x_1, x_2, x_3) \in D$ is the spatial coordinate;

2. the particle direction $\mathbf{\Omega} = (\Omega_1, \Omega_2, \Omega_3)$ is a point on the unit sphere S^2, with

$$\Omega_1 = \sin\theta\cos\phi, \quad \Omega_2 = \sin\theta\sin\phi \quad \text{and} \quad \Omega_3 = \cos\theta,$$

where θ is the polar angle and ϕ is the azimuthal angle;

3. the particle kinetic energy $E \in I$ is given by :

$$E = \frac{1}{2}m\,|\mathbf{v}|^2, \tag{5}$$

with $\mathbf{v}$ the particle velocity, m the particle mass and $|.|$ denoting the euclidian norm in $\mathbb{R}^3$. The particle speed is $v = |\mathbf{v}|$.

In all that follows, we assume that the velocity $\mathbf{v}$ belongs to a bound domain of $\mathbb{R}^3$. Thus

$$E \in I = [E_{min}, E_{max}], \ 0 < E_{min} < E_{max}. \tag{6}$$

The differential phase-space volume associated with the phase-space point $\mathbf{P}$ is given by :

$$dP = d\mathbf{x}d\mathbf{\Omega}dE,$$

where $d\mathbf{x} = dx_1 dx_2 dx_3$ and $d\mathbf{\Omega} = \sin\theta d\theta d\phi$. The goal of the transport theory being to determine the distribution of the particles in the domain D taking into account their movement and their interaction with the medium which they occupy, we define in this intention significant quantities (functions) which characterize this distribution.

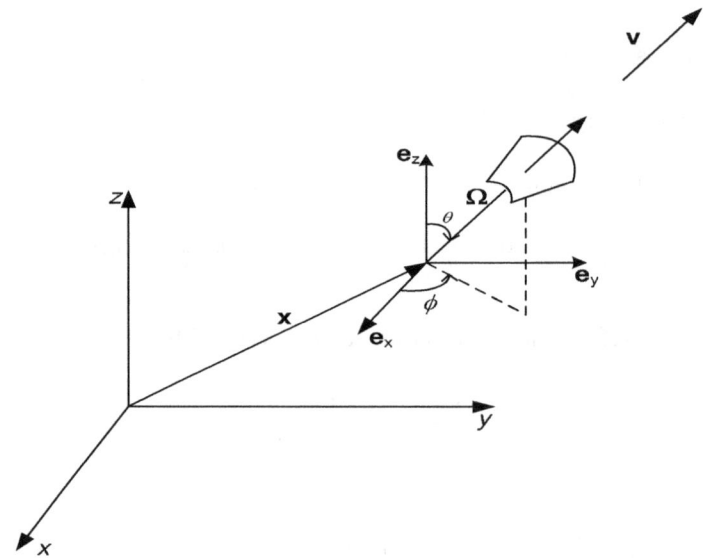

Figure 1: The position and direction variables characterizing a particle

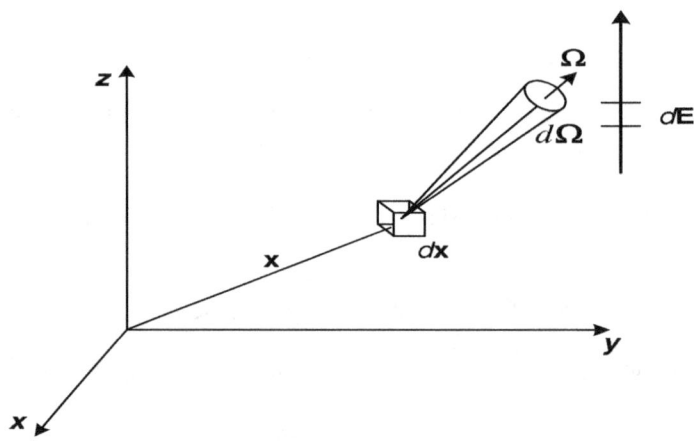

Figure 2: Differential phase-space volume $d\mathbf{P}$: element of volume $d\mathbf{x}$ around $\mathbf{x}$; solid angle $d\Omega$ around Ω and the element of distance dE about the energy E

1.2.1 Basic Definitions

In this section, we give definitions of some quantities most commonly used in particle transport.

Definition 1.2.1. *Phase-space particle density*

The fundamental quantity in modelling particle transport is the *phase-space particle density* also known as *angular particle density* and denoted by:

$$n(\mathbf{x}, \mathbf{\Omega}, E, t) \equiv n(\mathbf{P}, t). \tag{7}$$

It is defined as the probable number of particles at the position $\mathbf{x}$ travelling in the direction $\mathbf{\Omega}$ with and energy E at time t, per unit volume per unit solid angle per unit energy ($particles/(cm^3.Steradian.MeV)$). Hence, the quantity

$$n(\mathbf{P}, t)d\mathbf{P} \tag{8}$$

is the expected number of particles in the element of volume $d\mathbf{x}$ about $\mathbf{x}$ travelling in a direction in $d\mathbf{\Omega}$ about $\mathbf{\Omega}$ with an energy in dE about E at time t.

It is assumed that :

$$n(\cdot, \cdot, \cdot, t) \in L^1_{loc}(D \times S^2 \times I). \tag{9}$$

Definition 1.2.2. *Particle density*.

The total *particle density* or the *energy dependent particle density* $N(\mathbf{x}, E, t)$ is the expected number of particles at position $\mathbf{x}$, with energy E at time t. It is the integral of particle angular density over all direction :

$$N(\mathbf{x}, E, t) = \int_{S^2} n(\mathbf{x}, \mathbf{\Omega}, E, t)d\mathbf{\Omega}. \tag{10}$$

Hence

$$N(\mathbf{x}, E, t)d\mathbf{x}dE, \tag{11}$$

represents the expected number of particles in $d\mathbf{x}$ about $\mathbf{x}$ having energy ranging between E and $E + dE$ at time t

Definition 1.2.3. *Particle Angular Flux*.

The *particle angular flux* denoted by φ is the product of particle speed and particle angular density :

$$\varphi(\mathbf{P}, t) = v.n(\mathbf{P}, t), \tag{12}$$

11

where $v = |\mathbf{v}|$ is the particle speed.

The total number of particles passing at $\mathbf{x}$ through an area of one cm^2, perpendicular to $\mathbf{\Omega}$ per second with energy in dE about E and direction in $d\mathbf{\Omega}$ about $\mathbf{\Omega}$ at time t is given by :

$$\varphi(\mathbf{P}, t)d\mathbf{\Omega}dE \equiv v.n(\mathbf{P}, t)d\mathbf{\Omega}dE. \tag{13}$$

Definition 1.2.4. *Particle Scalar Flux*.

The particle *scalar flux* or *total particle flux* denoted by $\phi(\mathbf{x}, E, t)$ is the integral over all directions of the particle angular flux :

$$\phi(\mathbf{x}, E, t) = v.N(\mathbf{x}, E, t) = \int_{S^2} \varphi(\mathbf{P}, t)d\mathbf{\Omega}. \tag{14}$$

Definition 1.2.5. *Net Current Vector*.

The particle *current vector* denoted by $\mathbf{J}(\mathbf{x}, E, t)$ is the net number of particles crossing a surface element per unit energy in unit time. It is given by :

$$\mathbf{J}(\mathbf{x}, E, t) = v \int_{S^2} \mathbf{\Omega}.n(\mathbf{P}, t)d\mathbf{\Omega}. \tag{15}$$

Since $\varphi(\mathbf{P}, t) = v.n(\mathbf{P}, t)$, we have :

$$\mathbf{J}(\mathbf{x}, E, t) = \int_{S^2} \mathbf{\Omega}.\varphi(\mathbf{P}, t)d\mathbf{\Omega}. \tag{16}$$

It can be deduced from equations (14) and (16) that the particle scalar flux and current are, respectively, the zero's and the first moment of the particle angular flux.

1.2.2 Derivation of the Transport Equation

The particle angular density at $(\mathbf{x}, \mathbf{\Omega}, E)$ at time t results from the balance relation between the gains and the losses of particles in the domain D. The description of various mechanisms by which these phenomena appear leads to the particle transport equation.

Let us consider an arbitrary volume $V \subset \mathbb{R}^3$ occupied by particles in interaction with the medium. If the external forces which can change the trajectories of the particles in the domain V are neglected, the variation of the number of particles having direction in $d\mathbf{\Omega}$ about $\mathbf{\Omega}$ with an energy ranging between E and $E + dE$ in the course of time is essentially due to : their exit through the boundary ∂V of V, the collisions and the particle production sources in V. The resulting balance relation reads (Duderstadt, 1983) :

$$
\begin{pmatrix}
\text{Variation of} \\
\text{the number of} \\
\text{particles in } V
\end{pmatrix}
=
\begin{pmatrix}
\text{Variation due} \\
\text{to collisions} \\
\text{in } V
\end{pmatrix}
+
\begin{pmatrix}
\text{Exit through} \\
\text{boundary } \partial V \\
\text{of } V
\end{pmatrix}
$$

$$
+
\begin{pmatrix}
\text{Particle} \\
\text{production} \\
\text{sources in } V
\end{pmatrix}. \tag{17}
$$

It then follows that :

$$
\frac{\partial}{\partial t}\left(\int_V n(\mathbf{P},t)d\mathbf{x}d\mathbf{\Omega}dE\right) = -\int_{\partial V} \mathbf{\Omega}.\varphi(\mathbf{P},t).\mathbf{d\Gamma}d\mathbf{\Omega}dE + \int_V \left(\frac{\partial n}{\partial t}\right)_{coll} d\mathbf{x}d\mathbf{\Omega}dE
$$

$$
+\int_V s(\mathbf{P},t)d\mathbf{x}d\mathbf{\Omega}dE, \tag{18}
$$

where $s(\mathbf{P},t)$ denotes the density of the particle production sources in V, $\left(\frac{\partial n}{\partial t}\right)_{coll}$ describes the variations due to collisions and

$$
\mathbf{d\Gamma} = \mathbf{n} \cdot d\Gamma
$$

with $d\Gamma$ representing the element of surface and $\mathbf{n}$ denoting the outward unit normal vector.

By using the Gauss's theorem [37, 75]:

$$
\int_{\partial V} \mathbf{F}(\mathbf{x}) \cdot \mathbf{d\Gamma} = \int_V \nabla_x \cdot \mathbf{F}(\mathbf{x})dV, \tag{19}
$$

where $\nabla_x = \left(\frac{\partial}{\partial x_1}, \frac{\partial}{\partial x_2}, \frac{\partial}{\partial x_3}\right)$ denotes the gradient with respect to the spatial variable $\mathbf{x}$, we have :

$$
\int_{\partial V} (\mathbf{\Omega}.\varphi d\mathbf{\Omega}dE)\, \mathbf{d\Gamma} = \int_V (\nabla_x \cdot \mathbf{\Omega}\varphi)\, d\mathbf{x}d\mathbf{\Omega}dE = \int_V \mathbf{v} \cdot (\nabla_x n)\, d\mathbf{x}d\mathbf{\Omega}dE. \tag{20}
$$

Through substitution, equation (18) then reads :

$$
\int_V \left(\frac{\partial n}{\partial t} + \mathbf{v} \cdot \nabla_x n - \left(\frac{\partial n}{\partial t}\right)_{coll} - s\right) d\mathbf{x}d\mathbf{\Omega}dE = 0. \tag{21}
$$

Since (21) holds for any arbitrary volume V, we then obtain the following equation :

$$
\frac{\partial n}{\partial t} + \mathbf{v} \cdot \nabla_x n - \left(\frac{\partial n}{\partial t}\right)_{coll} - s = 0, \tag{22}
$$

13

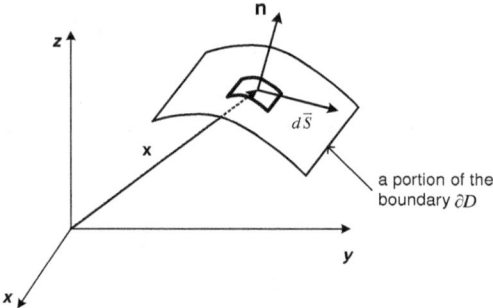

Figure 3: Definition of the outward unit normal vector **n** to the boundary ∂D at the point $\mathbf{x} \in \partial D$.

which describes the particle transport phenomena in several applications. It thus remains to clarify the variations due to collisions.

After a collision, the incident particles with velocity $\mathbf{v}'$ at position $\mathbf{x}$ is either absorbed, or scattered with a new velocity $\mathbf{v}$. The absorption phenomena is characterized by the captures and/or the fissions. Scattering, capture and fission strongly depend on the properties of the transport media. We define below, some of these properties.

Definition 1.2.6. *Microscopic interaction cross-section*

The microscopic interaction cross-section, denoted by $\hat{\Sigma}(\mathbf{x}, E)$, is the effective cross-sectional area of a target atom for a particular type of interaction seen at position $\mathbf{x}$, by a particle with energy E.

Definition 1.2.7. *Macroscopic cross-section*

The macroscopic or total cross-section, denoted by $\Sigma_t(\mathbf{x}, E)$, is the probability of particle interaction per unit distance travelled by particle with energy E at position $\mathbf{x}$. It is related to the microscopic interaction cross-section as follows:

$$\Sigma_t(\mathbf{x}, E) = \rho_a(\mathbf{x})\hat{\Sigma}(\mathbf{x}, E), \tag{23}$$

where $\rho_a(\mathbf{x})$ is the atomic density of the background medium.

The inverse of the macroscopic cross-section defines the *mean-free-path*. In the case of space-independent value of Σ_t, the mean-free-path is the average distance that a particle travelled between interactions.

Definition 1.2.8. *Macroscopic differential scattering cross-section*

14

The Macroscopic differential scattering cross-section is defined by :

$$\Sigma_s(\mathbf{x}, \mathbf{\Omega} \to \mathbf{\Omega}', E \to E') \equiv \Sigma_s(\mathbf{x}, \mathbf{v} \to \mathbf{v}')$$
$$= \Sigma_t(\mathbf{x}, E)c(\mathbf{x}, E)f(\mathbf{x}, \mathbf{\Omega} \to \mathbf{\Omega}', E \to E'), \qquad (24)$$

where :

- $c(\mathbf{x}, E)$ is the mean number of scattering particles emitted in a collision event experienced by an incident particle with energy E at $\mathbf{x}$;

- $f(\mathbf{x}, \mathbf{\Omega} \to \mathbf{\Omega}', E \to E') \equiv f(\mathbf{x}, \mathbf{v} \to \mathbf{v}')$: the probability that a particle with velocity $\mathbf{v}$ at position $\mathbf{x}$ is scattered with velocity $\mathbf{v}'$.

The expression

$$\Sigma_s(\mathbf{x}, \mathbf{\Omega} \to \mathbf{\Omega}', E \to E')ds d\mathbf{\Omega}' dE', \qquad (25)$$

is the expected number of scattering events experienced by a particle with initial energy E, and initial direction $\mathbf{\Omega}$, travels a differential distance ds, and scatters into a direction in $d\mathbf{\Omega}'$ about $\mathbf{\Omega}'$, with final energy in dE about E.

With these definitions, the scattering source is given by :

$$S_s(\mathbf{x}, \mathbf{\Omega}, E, t) = \int dE' \int d\mathbf{\Omega}' \Sigma_s(\mathbf{x}, \mathbf{v}' \to \mathbf{v})\varphi(\mathbf{x}, \mathbf{\Omega}', E', t). \qquad (26)$$

Let us denote by :

- $\Sigma_f(\mathbf{x}, E')$: the macroscopic fission cross-section. $\Sigma_f(\mathbf{x}, E')ds$ is the probability that a particle at $\mathbf{x}$ with energy E' will initiate a fission event while travelling a distance ds;

- $\chi(E)$: the energy dependent spectrum of particles emitted by fission. ;

- $v(\mathbf{x}, E)$: the number of particle with energy E, produced per fission at $\mathbf{x}$.

The fission source is given by :

$$S_f(\mathbf{x}, \mathbf{\Omega}, E, t) = \frac{\chi(E)}{4\pi} \int dE' v(\mathbf{x}, E) \Sigma_f(\mathbf{x}, E') \int_{S^2} \varphi(\mathbf{x}, \mathbf{\Omega}', E', t) d\mathbf{\Omega}'. \qquad (27)$$

Therefore the variations due to collisions read [37, 34, 75] :

$$\left(\frac{\partial n}{\partial t}\right)_{coll} = v\Sigma_t(\mathbf{x}, E)n(\mathbf{P}, t) - S_s(\mathbf{P}, t) - S_f(\mathbf{P}, t), \qquad (28)$$

15

and equation (22) becomes :

$$\frac{\partial n}{\partial t}(\mathbf{P}, t) = -\mathbf{v} \cdot \nabla_x n(\mathbf{P}, t) - v\Sigma_t(\mathbf{x}, E)n(\mathbf{P}, t)$$

$$+ \int dE' \int d\mathbf{\Omega}' \Sigma_s(\mathbf{x}, \mathbf{v}' \to \mathbf{v})\varphi(\mathbf{x}, \mathbf{\Omega}', E', t) \qquad (29)$$

$$+ \frac{\chi(E)}{4\pi} \int dE' v(\mathbf{x}, E)\Sigma_f(\mathbf{x}, E') \int_{S^2} \varphi(\mathbf{x}, \mathbf{\Omega}', E', t) d\mathbf{\Omega}' + s(\mathbf{P}, t).$$

The equation (29) is equivalent to:

$$\frac{1}{v}\frac{\partial \varphi}{\partial t}(\mathbf{P}, t) = -\mathbf{\Omega} \cdot \nabla_x \varphi(\mathbf{P}, t) - \Sigma_t(\mathbf{x}, E)\varphi(\mathbf{P}, t)$$

$$+ \int dE' \int d\mathbf{\Omega}' \Sigma_s(\mathbf{x}, \mathbf{v}' \to \mathbf{v})\varphi(\mathbf{x}, \mathbf{\Omega}', E', t) \qquad (30)$$

$$+ \frac{\chi(E)}{4\pi} \int dE' v(\mathbf{x}, E)\Sigma_f(\mathbf{x}, E') \int_{S^2} \varphi(\mathbf{x}, \mathbf{\Omega}', E', t) d\mathbf{\Omega}' + s(\mathbf{P}, t).$$

The relations (29) and (30) define the time-dependent first order neutral particle transport equations.

The equation (30) can be recast as follows :

$$\frac{1}{|\mathbf{v}|}\frac{\partial \varphi}{\partial t}(\mathbf{P}, t) + \mathbf{\Omega} \cdot \nabla_x \varphi + \Sigma_t(\mathbf{x}, E)\varphi = \int_B k(\mathbf{x}, \mathbf{\Omega}, \mathbf{\Omega}', E, E')\varphi(\mathbf{x}, \mathbf{\Omega}', E', t)dE'd\mathbf{\Omega}'$$

$$+ s(\mathbf{P}, t), \qquad (31)$$

where $\mathbf{B} = S^2 \times [E_{min}, E_{max}]$ and,

$$k(\mathbf{x}, \mathbf{\Omega}, \mathbf{\Omega}', E, E') = \Sigma_s(\mathbf{x}, \mathbf{\Omega}' \to \mathbf{\Omega}, E' \to E) + \frac{\chi(E)}{4\pi}v(\mathbf{x}, E')\sigma_f(\mathbf{x}, E'), \qquad (32)$$

denotes the *collision kernel* or the *transfer kernel*, which describes the expected distribution of particles emerging from scattering events, fissions, etc...

The Cauchy problem for the particle transport equation consists to determine the unknown function $\varphi(\mathbf{x}, \mathbf{\Omega}, E, t)$ verifying equation (31) and satisfying the initial condition

$$\varphi(\mathbf{x}, \mathbf{\Omega}, E, 0) = \varphi_0(\mathbf{x}, \mathbf{\Omega}, E) \qquad (33)$$

and some boundary conditions at the boundary ∂D of D. The function $\varphi_0(\mathbf{x}, \mathbf{\Omega}, E)$ denoting the angular flux at the initial time $t = 0$.

1.2.3 Boundary Conditions

The angular flux $\varphi(\mathbf{P}, t)$ has to satisfy the boundary conditions at various surface to give a unique solution to the particle transport equation (31) subject to the initial condition (33).

Let S be the enclosing surface of the system ($S \equiv \partial D$). The most common surfaces that can occur over the surface S are:

1. a bare surface, denoted by S_b,

2. a surface with a source denoted by S_s,

3. a perfect reflecting surface, denoted by S_{pr}.

These surfaces define the boundary conditions used in conjunction with the first order transport equation (30) as follows :

A bare surface or vacuum surface: No particle enter the the system across the bare surface and particles crossing a bare surface in an outward direction leave the system forever. Thus

$$\varphi(\mathbf{x}, \mathbf{\Omega}, E, t) = 0, \ \mathbf{\Omega} \cdot \mathbf{n}(\mathbf{x}) < 0, \tag{34}$$

where $\mathbf{n}(\mathbf{x})$ is the outward normal at $\mathbf{x}$ to S_b.

A surface with source: Particles enter the system from external source across S_s and particles crossing a surface with source in an outward direction leave the system forever. The incoming particle flux must be specified. Thus

$$\varphi(\mathbf{x}, \mathbf{\Omega}, E, t) = g(\mathbf{x}, \mathbf{\Omega}, E, t), \ \mathbf{\Omega} \cdot \mathbf{n}(\mathbf{x}) < 0. \tag{35}$$

A bare surface is a particular case of surface with source, where the incoming angular flux $g(\mathbf{x}, \mathbf{\Omega}, E, t) = 0$.

A perfect reflector surface: At a point $\mathbf{x}$ on a perfectly reflection surface, the incident particle beam with direction $\mathbf{\Omega}$ and the reflected beam with direction $\mathbf{\Omega}^*$ are equally inclined to the normal $\mathbf{n}(\mathbf{x})$. The boundary condition is :

$$\varphi(\mathbf{x}, \mathbf{\Omega}, E, t) = \varphi(\dot{\mathbf{x}}, \mathbf{\Omega}^*, E, t), \tag{36}$$

where

$$\mathbf{\Omega}^* = \mathbf{\Omega} - 2(\mathbf{\Omega} \cdot \mathbf{n}(\mathbf{x}))\mathbf{n}(\mathbf{x}), \ \mathbf{x} \in S_{pr} \ \text{and} \ \mathbf{\Omega} \cdot \mathbf{n}(\mathbf{x}) \neq 0. \tag{37}$$

In addition to these boundary conditions, internal boundary or interface conditions occur for system composed of regions presenting different physical compositions. The angular flux must be a continuous of $\mathbf{x}$ for all directions $\mathbf{\Omega}$ at the internal interfaces except for those directions which are along the interfaces $\mathbf{\Omega} \cdot \mathbf{n}(\mathbf{x}) = 0$.

1.3 Properties of the transport equation

We present in this section, some properties of the transport operator. The existence and uniqueness theorem for the transport problem is given

By using the notation of [34], we set:

$$u(\mathbf{x}, \mathbf{v}, t) \equiv \frac{m}{v} \varphi(\mathbf{x}, \mathbf{\Omega}, E, t) \tag{38}$$

$$\sigma(\mathbf{x}, \mathbf{v}) \equiv v \Sigma(\mathbf{x}, E) \tag{39}$$

$$f(\mathbf{x}, \mathbf{v}, \mathbf{v}') \equiv m \frac{v'}{v} \left[\Sigma_s(\mathbf{x}, \mathbf{v} \to \mathbf{v}') + \frac{\chi(E)}{4\pi} \nu(\mathbf{x}, E') \Sigma_f(\mathbf{x}, E') \right] \tag{40}$$

$$q(\mathbf{x}, \mathbf{v}, t) \equiv \frac{m}{v} s(\mathbf{x}, \mathbf{\Omega}, E, t) \tag{41}$$

Let

$$X = D \times V, \tag{42}$$

where $V \subset \mathbb{R}^3$ is the velocity space. We set

$$\Gamma = \partial D \times V, \tag{43}$$

$$\Gamma^+ = \{(\mathbf{x}, \mathbf{v}) \in \Gamma, \ \mathbf{v} \cdot \mathbf{n}(\mathbf{x}) > 0\}, \tag{44}$$

$$\Gamma^- = \{(\mathbf{x}, \mathbf{v}) \in \Gamma, \ \mathbf{v} \cdot \mathbf{n}(\mathbf{x}) < 0\}, \tag{45}$$

$$\Gamma^0 = \{(\mathbf{x}, \mathbf{v}) \in \Gamma, \ \mathbf{v} \cdot \mathbf{n}(\mathbf{x}) = 0\}. \tag{46}$$

It is assumed that the measure of the set $\Gamma \setminus (\Gamma^+ \cup \Gamma^-)$ is null with respect to the $dsdv$-measure on Γ, where ds is the measure induced on ∂D by the Lebesgue measure on $\mathbb{R}^3$ [23].

We consider the following Cauchy problem for the particle transport: find a function $u(., ., t) \in L^p(X)$, $p \in [1, +\infty[$, such that :

$$\begin{cases} \frac{\partial u}{\partial t} + Lu = Ku + q(\mathbf{x}, \mathbf{v}, t), & \text{in } X \times]0, \tau_1[, \\ u(\mathbf{x}, \mathbf{v}, t) = g(\mathbf{x}, \mathbf{v}, t) & (\mathbf{x}, \mathbf{v}, t) \in \Gamma^- \times]0, \tau_1[, \\ u(\mathbf{x}, \mathbf{v}, 0) = u^0(\mathbf{x}, \mathbf{v}), \end{cases} \tag{47}$$

where $L^p(X)$ denotes the space of $d\mathbf{x}d\mathbf{v}$−measurable functions f on X satisfying

$$\|f\|_{L^p(X)} = \left(\int_X |f(x, v)|^p dx dv \right)^{\frac{1}{p}} < +\infty \tag{48}$$

and

- u^0 and g are given functions;

18

- K is the integral scattering operator defined by

$$Ku(\mathbf{x}, \mathbf{v}, t) = \int_V f(\mathbf{x}, \mathbf{v}', \mathbf{v}) u(\mathbf{x}, \mathbf{v}', t) d\mathbf{v}', \tag{49}$$

with f a given nonnegative measurable function on S^2;

- L is the streaming operator defined by

$$Lu(\mathbf{x}, \mathbf{v}, t) = \mathbf{\Omega} \cdot \nabla_x u(\mathbf{x}, \mathbf{v}, t) + \sigma(\mathbf{x}, \mathbf{v}) u(\mathbf{x}, \mathbf{v}, t), \tag{50}$$

with σ a given nonnegative function defined on X.

In all that follows, $L^p(B, \nu)$ ($1 \leq p < +\infty1$) denotes the space of ν-measurable on real functions f on B such that

$$\|f\|_{L^p(B,\nu)} = \left(\int_B |f|^p d\nu \right)^{\frac{1}{p}} < +\infty, \tag{51}$$

where B is a locally compact space and ν is a positive measure on B.

To analyze the problem (47), the following spaces are introduced :

$$L^p(X \times]0, \tau_1[) = L^p(X \times]0, \tau_1[, \; d\mathbf{x} d\mathbf{v} dt); \tag{52}$$

$$L^p(\Gamma^{\pm} \times]0, \tau_1[) = L^p(\Gamma^{\pm} \times]0, \tau_1[, \; |\mathbf{v} \cdot \mathbf{n}(\mathbf{x})| ds d\mathbf{v} dt) \tag{53}$$

$$L^p(\Gamma^{\pm}) = L^p(\Gamma^{\pm}, \; |\mathbf{v} \cdot \mathbf{n}(\mathbf{x})| ds d\mathbf{v}) \tag{54}$$

$$\mathcal{W}_p(X \times]0, \tau_1[) = \{u \in L^p(X \times]0, \tau_1[), \; \frac{\partial u}{\partial t} + \mathbf{v} \cdot \nabla_x u \in L^p(X \times]0, \tau_1[)$$
$$u(., t) \in L^p(X), \; u|_{\Gamma^- \times (0, \tau_1)} \in L^p(\Gamma^- \times (0, \tau_1))\} \tag{55}$$

$$W^p(X) = \{u \in L^p(X), \; \mathbf{v} \cdot \nabla_x u \in L^p(X)\}. \tag{56}$$

Let us set

$$d\xi = |\mathbf{v} \cdot \mathbf{n}(\mathbf{x})| \min(\tau(\mathbf{x}, \mathbf{v}), C) ds d\mathbf{v},$$

$$d\xi_r = |\mathbf{v} \cdot \mathbf{n}(\mathbf{x})| \min(\tau(\mathbf{x}, \mathbf{v}), \tau_1 - t, C) ds d\mathbf{v} dt,$$

where C is a positive constant and

$$\tau(\mathbf{x}, \mathbf{v}) = \begin{cases} \inf\{t > 0, \; \mathbf{x} + \mathbf{v}t \notin D\} \; if \; (\mathbf{x}, \mathbf{v}) \in \Gamma^- \\ \inf\{t > 0, \; \mathbf{x} - \mathbf{v}t \notin D\} \; if \; (\mathbf{x}, \mathbf{v}) \in \Gamma^+ \end{cases}.$$

Functions in $W^p(X)$ have trace on $\Gamma^{\pm}$ in $L^p(\Gamma^{\pm}, d\xi)$ (see [29, 34]).

The space W_p is equipped with norm

$$\|u\|_{W^p} = \left(\|u\|_{L^p(X)}^p + \|\mathbf{v} \cdot \nabla_x u\|_{L^p(X)}^p \right)^{\frac{1}{p}}. \tag{57}$$

19

1.3.1 Existence and Uniqueness Results for the Time Dependent Transport Equation

We set in the following the existence and uniqueness theorem for the solution of the problem (47).

We assume that $g(\mathbf{x}, \mathbf{v}, t) \equiv 0$. We have the following results [34, 29]:

Theorem 1.3.1. *Existence and uniqueness theorem for problem (47)*

We assume that the data of problem (47) satisfy:

1. *$\sigma \in L^{\infty}(X)$, $\sigma \geq 0$ and $f \in L^{\infty}(X)$,*

2. *The transfer kernel f is $d\mathbf{v}$-measurable, $d\mathbf{v}'$-measurable and there exits nonnegative constants M_a and M_b such that:*

$$\int_V f(\mathbf{x}, \mathbf{v}', \mathbf{v}) d\mathbf{v} \leq M_a, \quad \forall (\mathbf{x}, \mathbf{v}) \in X, \tag{58}$$

$$\int_V f(\mathbf{x}, \mathbf{v}', \mathbf{v}) d\mathbf{v}' \leq M_b, \quad \forall (\mathbf{x}, \mathbf{v}) \in X \tag{59}$$

3. *$q \in L^p(X \times [0, \tau_1])$, $p \in [1, +\infty[$,*

4. *$u^0 \in L^p(X)$.*

Then the problem (47) admits unique solution $u \in \mathcal{W}_p(X \times]0, \tau_1[)$. The solution u satisfies:

$$u \in \mathcal{C}^0([0, \tau_1]; L^p(X)). \tag{60}$$

Additionally, if function u^0 and q satisfy :

$$\mathbf{v} \cdot \nabla_x u^0 \in L^p(X) \quad and \quad u^0|_{\Gamma^-} = 0, \tag{61}$$

$$q \in \mathcal{C}^1([0, \tau_1]; L^p(X)), \tag{62}$$

then the solution u of problem (47) satisfies :

$$u \in \mathcal{C}^1([0, \tau_1]; L^p(X)), \quad \mathbf{v} \cdot \nabla_x u \in \mathcal{C}^0([0, \tau_1]; L^p(X)), \quad and \quad u(t)|_{\Gamma^-} = 0, \forall t \in [0, \tau_1].$$

If $q \geq 0$ and $u^0 \geq 0$, then $u \geq 0$.

Proof. See [34] (Chapter XXI) for a detailed proof. □

In the case where

$$g(\mathbf{x}, \mathbf{v}, t) \neq 0,$$

assuming that

$$g \in L^p(\Gamma^- \times (0, \tau_1), d\xi_r),$$

It follows by using the properties of the trace map [34] that there exists a function $\tilde{u} \in \mathcal{W}_p(X \times]0, \tau_1[)$ of which g is the trace on $\Gamma^- \times (0, \tau_1)$. Therefore, setting

$$w = u - \tilde{u},$$

the problem (47) is equivalent to

$$\begin{cases} \frac{\partial w}{\partial t} + Lw = Kw + \tilde{q}(\mathbf{x}, \mathbf{v}, t), & \text{in } X \times]0, \tau_1[, \\ w(\mathbf{x}, \mathbf{v}, t)_{|\Gamma^- \times]0, \tau_1[} = 0 & (63) \\ w(\mathbf{x}, \mathbf{v}, 0) = w^0(\mathbf{x}, \mathbf{v}), \end{cases}$$

where $\tilde{q} = q - \frac{\partial \tilde{u}}{\partial t} - L\tilde{u} - K\tilde{u}$ and $w^0 = u^0 - \tilde{u}$ are such that $\tilde{q} \in L^p(X \times [0, \tau_1])$, $p \in [1, +\infty[$ and $w^0 \in L^p(X)$.

It follows then from Theorem 1.3.1, the existence and uniqueness of the solution of problem (63) in $\mathcal{W}_p$. Thus the problem (47) admits a unique solution $u \in \mathcal{W}_p$ satisfying

$$u \in \mathcal{C}^0([0, \tau_1]; L^p(X)). \tag{64}$$

Moreover, if the functions g, q and u^0 are nonnegative, then the solution u of (47) is nonnegative [34].

1.3.2 Existence and Uniqueness Results for Time Independent Transport Equation

By removing the time dependency of the flux and assuming that the source function q and the boundary conditions are independent of time t, the time independent or the steady state form of the particle transport problem (47) is obtained and reads:

$$\begin{cases} -Tu(\mathbf{x}, \mathbf{v}) = q(\mathbf{x}, \mathbf{v}) & a.e. \ (\mathbf{x}, \mathbf{v}) \in X, \\ u(\mathbf{x}, \mathbf{v}) = u_b(\mathbf{x}, \mathbf{v}) & a.e. \ (\mathbf{x}, \mathbf{v}) \in \Gamma^-, \end{cases} \tag{65}$$

where T is the transport operator defined by:

$$T = -L + K, \tag{66}$$

21

with the operators L and K defined respectively by (49) and (50).

We have the following existence results [34, 29]:

Theorem 1.3.2. *Existence and uniqueness theorem for problem (65)*

We assume that the data of problem (65) satisfy:

1. $\sigma \in L^\infty(X)$, $\sigma \geq 0$ and $f \in L^\infty(X)$,

2. The transfer kernel f is $d\mathbf{v}$-measurable, $d\mathbf{v}'$-measurable and there exits nonnegative constants M_a and M_b such that:

$$\int_V f(\mathbf{x}, \mathbf{v}', \mathbf{v}) d\mathbf{v} \leq M_a, \quad \forall (\mathbf{x}, \mathbf{v}) \in X, \tag{67}$$

$$\int_V f(\mathbf{x}, \mathbf{v}', \mathbf{v}) d\mathbf{v}' \leq M_b, \quad \forall (\mathbf{x}, \mathbf{v}) \in X \tag{68}$$

3. There exists a positive constant α such that the functions σ and f satisfy a.e $(\mathbf{x}, \mathbf{v}) \in X$

$$\sigma(\mathbf{x}, \mathbf{v}) - \int_V f(\mathbf{x}, \mathbf{v}, \mathbf{v}') d\mathbf{v}' \geq \alpha, \tag{69}$$

$$\sigma(\mathbf{x}, \mathbf{v}) - \int_V f(\mathbf{x}, \mathbf{v}', \mathbf{v}) d\mathbf{v}' \geq \alpha. \tag{70}$$

4. $q \in L^p(X)$, $p \in [1, +\infty[$,

5. $u_b \in L^p(\Gamma^-, d\xi)$.

Then for $1 \leq p < +\infty$, the problem (65) admits unique solution $u \in D(L)$, where $D(L)$ is the domain of the operator L defined by:

$$D(L) = \{u \in W^p(X); u|_{\Gamma^-} = u_b\}.$$

The solution u satisfies:

$$\|u\|_{L^p(X)} \leq C_1 \left(\|q\|_{L^p(X)} + \|u_b\|_{L^p(\Gamma^-)} \right), \tag{71}$$

where C_1 is a positive constant independent of u.

Theorem 1.3.3. *Let $p = 2$. Under the assumptions of Theorem 1.3.2, we have the following bound for the solution of problem (65) [29, 23, 34]:*

$$\|u\|_{W^2(X)} \leq C_1 \left(\|q\|_{L^2(X)} + \|u_b\|_{L^2(\Gamma^-)} \right), \tag{72}$$

where C_1 is a positive constant independent of u.

For the case $p = +\infty$, we have the following existence results [34, 29]:

Theorem 1.3.4. *Under the assumptions 1., 2. and 3. of Theorem 1.3.2, if*

- $\sigma \geq \sigma_0 > 0$,

- *the functions σ and f satisfy for all $(\mathbf{x}, \mathbf{v}) \in X$*

$$\int_V f(\mathbf{x}, \mathbf{v}', \mathbf{v}) d\mathbf{v}' \geq \beta \sigma(\mathbf{x}, \mathbf{v}), \ \ 0 \leq \beta < 1, \tag{73}$$

- $q \in L^\infty(X)$,

- $u_b \in L^\infty(\Gamma^-)$,

then the problem (65) admits unique solution $u \in L^\infty(X)$, which satisfies:

$$\|u\|_{L^\infty(X)} \leq C_1 \left(\|q\|_{L^\infty(X)} + \|u_b\|_{L^\infty(\Gamma^-)} \right), \tag{74}$$

where C_1 is a positive constant independent of u.

1.4   Other Forms of the Transport Equation

We present in this section some alternative form of the transport equation generally used for the numerical approximation. We only consider the steady state case. In the following, we set :

$$\sigma(\mathbf{x}, \mathbf{v}) \equiv \sigma(\mathbf{x}, E), \ \ \varphi(\mathbf{x}, \mathbf{v}) \equiv \varphi(\mathbf{x}, \mathbf{\Omega}, E), \ \ f(\mathbf{x}, \mathbf{v}', \mathbf{v}) \equiv f(\mathbf{x}, \mathbf{\Omega}', \mathbf{\Omega}, E', E). \tag{75}$$

1.4.1   The Integral Equation

The integral form of the stationary transport equation is obtained by inverting the streaming-collision operator L (see [34]). It reads :

$$
\begin{aligned}
\varphi(\mathbf{x}, \mathbf{\Omega}, E) &= \varphi(\mathbf{x} - \tau\mathbf{\Omega}, \mathbf{\Omega}, E) \exp\left(- \int_0^\tau \sigma(\mathbf{x} - \tau'\mathbf{\Omega}, E) d\tau' \right) \\
&+ \int_0^\tau Q(\mathbf{x} - s\mathbf{\Omega}, \mathbf{\Omega}, E) \exp\left[-\alpha(\mathbf{x} - s\mathbf{\Omega}, \mathbf{x}) \right] ds
\end{aligned}
\tag{76}
$$

where $\alpha(\mathbf{x}, \mathbf{x}', E)$ is the optical distance between $\mathbf{x}$ and $\mathbf{x}'$ defined by:

$$\alpha(\mathbf{x}, \mathbf{x}', E) = \int_0^{|\mathbf{x} - \mathbf{x}'|} \sigma \left(x + s \frac{\mathbf{x} - \mathbf{x}'}{|\mathbf{x} - \mathbf{x}'|} \right) ds \tag{77}$$

and

$$Q(\mathbf{x}, \mathbf{\Omega}, E) = K\varphi(\mathbf{x}, \mathbf{\Omega}, E) + q(\mathbf{x}, \mathbf{\Omega}, E). \tag{78}$$

23

1.4.2 Even Parity and Odd Parity forms

The transport equations for the directions Ω and $-\Omega$ read :

$$\Omega \cdot \nabla_x \varphi(\mathbf{x}, \Omega, E) + \sigma(\mathbf{x}, E)\varphi(\mathbf{x}, \Omega, E) = K\varphi(\mathbf{x}, \Omega, E) + q(\mathbf{x}, \Omega, E) \quad (79)$$

$$-\Omega \cdot \nabla_x \varphi(\mathbf{x}, -\Omega, E) + \sigma(\mathbf{x}, E)\varphi(\mathbf{x}, -\Omega, E) = K^-\varphi(\mathbf{x}, \Omega, E) + q(\mathbf{x}, -\Omega, E) \quad (80)$$

where the operator K^- is given by :

$$K^-\varphi(\mathbf{x}, \Omega, E) = \int_I \int_{S^2} f(\mathbf{x}, \Omega', -\Omega, E'E)\varphi(\mathbf{x}, \Omega', E')dE'd\Omega'$$

sum equation (80) with (79) and substract ing and substract (80) from (79) yield the following system of equations [34, 2, 82, 61]:

$$\Omega \cdot \nabla_x \varphi^o(\mathbf{x}, \Omega, E) + \sigma(\mathbf{x}, E)\varphi^e(\mathbf{x}, \Omega, E) = K^e\varphi(\mathbf{x}, \Omega, E) + q^e(\mathbf{x}, \Omega, E) \quad (81)$$

$$\Omega \cdot \nabla_x \varphi^e(\mathbf{x}, \Omega, E) + \sigma(\mathbf{x}, E)\varphi^o(\mathbf{x}, \Omega, E) = K^o\varphi(\mathbf{x}, \Omega, E) + q^o(\mathbf{x}, \Omega, E) \quad (82)$$

where for a given function $g(\mathbf{x}, \Omega, E)$, g^e and g^o are the even and odd angular flux defined respectively by the following Vladimirov transformation (see [82]) :

$$g^e(\mathbf{x}, \Omega, E) = \frac{1}{2}[g(\mathbf{x}, \Omega, E) + (\mathbf{x}, -\Omega, E)] \quad (83)$$

$$g^o(\mathbf{x}, \Omega, E) = \frac{1}{2}[g(\mathbf{x}, \Omega, E) - (\mathbf{x}, -\Omega, E)] \quad (84)$$

and the operators K^e and K^o are given by :

$$K^e = \frac{1}{2}(K + K^-) \quad (85)$$

$$K^o = \frac{1}{2}(K - K^-). \quad (86)$$

We have

$$\varphi = \varphi^e + \varphi^o$$

and the functions φ^o and φ^e satisfy :

$$K^e\varphi^o = K^o\varphi^e = 0.$$

The following system of parity equations is then obtained from equations (81) and (82) :

$$\Omega \cdot \nabla_x \varphi^o + \sigma\varphi^e = K^e\varphi^e + q^e \quad (87)$$

$$\Omega \cdot \nabla_x \varphi^e + \sigma\varphi^o = K^o\varphi^o + q^o. \quad (88)$$

24

Solving for φ^o in (88), we get

$$\varphi^o = -(\sigma I - K^o)^{-1}\mathbf{\Omega}\cdot\nabla_x\varphi^e + (\sigma I - K^o)^{-1}q^o. \qquad (89)$$

Substituting (89) into (87) yields the following self-adjoint even-parity form of the transport equation

$$-\mathbf{\Omega}\cdot\nabla_x(\sigma I - K^o)^{-1}\mathbf{\Omega}\cdot\nabla_x\varphi^e + \sigma\varphi^e = K^e\varphi^e + q^e - \mathbf{\Omega}\cdot\nabla_x(\sigma I - K^o)^{-1}q^o. \qquad (90)$$

The vacuum boundary condition for the even-parity transport equation is specified as

$$\begin{cases} \varphi^e + (\sigma I - K^o)^{-1}\left(q^o - \mathbf{\Omega}\cdot\nabla_x\varphi^e\right) = 0 \\ \varphi^e - (\sigma I - K^o)^{-1}\left(q^o - \mathbf{\Omega}\cdot\nabla_x\varphi^e\right) = 0 \end{cases} \qquad (91)$$

Other types of boundary conditions are specified in [2, 75]. The self-adjoint odd-parity equation can be obtained in a similar way.

Important quantities of transport theory such as the total flux and the neutron current vector can be directly obtained from the computation of even and odd angular fluxes. For instance, we have:

$$\phi(\mathbf{x}, E) = \int_{S^2} \varphi(\mathbf{x}, \mathbf{\Omega}, E)d\mathbf{\Omega} = \int_{S^2} \varphi^e(\mathbf{x}, \mathbf{\Omega}, E)d\mathbf{\Omega} \qquad (92)$$

$$\mathbf{J}(\mathbf{x}, E) = \int_{S^2} \mathbf{\Omega}\varphi(\mathbf{x}, \mathbf{\Omega}, E)d\mathbf{\Omega} = \int_{S^2} \mathbf{\Omega}\varphi^o(\mathbf{x}, \mathbf{\Omega}, E)d\mathbf{\Omega} \qquad (93)$$

Substituting (89) into (93), the neutron current vector can be expressed in term of the even angular flux as:

$$\mathbf{J}(\mathbf{x}, E) = \int_{S^2} \mathbf{\Omega}(\sigma I - K)^{-1}[q - \mathbf{\Omega}\cdot\nabla_x\varphi](\mathbf{x}, \mathbf{\Omega}, E)d\mathbf{\Omega}. \qquad (94)$$

1.4.3 Self-Adjoint Angular Flux (SAAF) form

For the derivation of the SAAF form of the transport equation, we consider the equation (79). Solving for φ in (79), we get :

$$\varphi = -(\sigma I - K)^{-1}\mathbf{\Omega}\cdot\nabla_x\varphi + (\sigma I - K)^{-1}q. \qquad (95)$$

Substituting (95) into the gradient term of (79) gives the following SAAF form of the transport equation [64] :

$$-\mathbf{\Omega}\cdot\nabla_x(\sigma I - K)^{-1}\mathbf{\Omega}\cdot\nabla_x\varphi + (\sigma I - K)\varphi = q - \mathbf{\Omega}\cdot\nabla_x(\sigma I - K)^{-1}q. \qquad (96)$$

The boundary conditions consist of standard influx boundary conditions, along with an additional outflux boundary [11] stated as follows [64] :

$$[\varphi + (\sigma I - K)^{-1}\mathbf{\Omega}\cdot\nabla_x\varphi]\,(\mathbf{x}, \mathbf{\Omega}, E) = (\sigma I - K)^{-1}q(\mathbf{x}, \mathbf{\Omega}, E), \quad \mathbf{n}(\mathbf{x})\cdot\mathbf{\Omega} > 0. \qquad (97)$$

1.4.4 Diffusion Approximation

The diffusion approximation is one of the simplest means to solve the neutron transport equation. It is based on the Fick's law [29, 51] which suggests that the neutron current is proportional to the gradient of the total flux :

$$\mathbf{J}(\mathbf{x}, E) \;=\; -D(\mathbf{x}, E)\nabla\phi(\mathbf{x}, E), \tag{98}$$

where $D(\mathbf{x})$ is the diffusion coefficient which is approximated by $\frac{1}{\sigma(\mathbf{x},E)}$. The relation (99) is valid under the assumption of slow spatial variations. That is (see [41]) : a) weak geometrical heterogeneity; b) the cross-sections are isotropic and the scattering cross-section dominates the absorbtion cross-section; c) the position is not to closed to concentrated sources and the interface of different physical medias.

Integrating the transport equation (79) over the angular domain and using the Fick's law (99) yields the following diffusion approximation of the transport equation :

$$[-\nabla \cdot D(\mathbf{x}, E)\nabla + \sigma(\mathbf{x}, E)]\phi(\mathbf{x}, E) = \int_I f(\mathbf{x}, E, E')\phi(\mathbf{x}, E')dE' + q_0(\mathbf{x}, E), \tag{99}$$

where $q_0(\mathbf{x}, E) = \int_{S^2} q(\mathbf{x}, \mathbf{\Omega}, E)d\mathbf{\Omega}$.

1.5 Summary

Physical derivation of the neutron transport equation has been given in this chapter. The existence and uniqueness results were presented for the time-dependent and the time-independent equations. Additionally, some alternative forms of this equation such as integral form, second order forms and the diffusion approximation, which can reduce the complexity of the initial first order integro-differential transport equations in certain circumstances were discussed.

Chapter 2

Numerical Approximation of the Neutron Transport Equation

In this chapter we give an overview of deterministic methods for the solution of the neutron transport equation. Discretization methods for each of its variables are briefly discussed and the iterative approaches for solving the transport equations are presented.

2.1 Introduction

The neutron transport equation has been derived in Chapter 1. It is an integro-differential equation with seven independent variables and cannot be solved analytically in most cases of interest. The solution of this equation has to be approximated numerically. Deterministic methods are widely used for the approximation of the angular flux in the phase-space [47]. The first step of deterministic methods consists in discretizing the transport equation with respect to each of its variables : The energy variable is often discretized by multigroup approximation, in which the energy range is divided into energy groups and the interaction cross sections are approximated by histograms in energy, each histogram having one value within each energy group [49, 34]. This yields the multigroup time-dependent or time-independent according to the equation initially considered. The treatment of the time-dependent transport equation can be done by using varieties of methods such as finite difference and variational methods [2]. Finite-differencing the derivative in time is the widely used approach. The energy and time discretization lead to a coupled system of steady state single energy equations which only depend on the spatial and the angular variables. The angular discretization is often accomplished by

27

the integral method, discrete ordinates method (S_N) and the expansion of the angular flux in terms of angular basis functions such as spherical harmonics method (P_N), Walsh functions, wavelet functions and others (see [49, 34, 27, 26, 28, 78] and the references therein). The finite difference methods, finite element methods, nodal methods and the method of characteristics are usually used for the spatial discretization (see [2, 3, 4, 18, 19, 49, 48, 29, 34, 23, 61, 60]). The second step of deterministic methods consists in solving the system of algebraic equations resulting from the discretization of the transport problem. Due to the number of independent variables of the transport problem, this system is very large and thus difficult to invert directly. The solution strategy for solving the resulting system of equations has then focussed on iterative methods [47].

In this Chapter, we consider an open, bounded and convex domain $D \subset \mathbb{R}^3$ in interaction with neutral particles. At time t, the particle angular flux $\varphi(\mathbf{x}, \mathbf{\Omega}, E, t)$ depends on the position $\mathbf{x} \in D$, the direction $\mathbf{\Omega} \in S^2$ (the unit sphere in $\mathbb{R}^3$) and the energy E. It verifies the following unsteady state (time dependent) linear neutron transport equation (NTE) (see [2, 34]):

$$\left[\frac{1}{v} \frac{\partial}{\partial t} + \mathbf{\Omega}.\nabla_x + \Sigma(\mathbf{x}, E) \right] \varphi(\mathbf{x}, \mathbf{\Omega}, E, t) =$$

$$\int dE' \int_{S^2} d\mathbf{\Omega}' \Sigma_s(\mathbf{x}, \mathbf{\Omega}' \to \mathbf{\Omega}, E' \to E) \varphi(\mathbf{x}, \mathbf{\Omega}', E', t) + S(\mathbf{x}, \mathbf{\Omega}, E, t), \qquad (1)$$

where v is the speed, $t \in]0, \tau[$ is the time, $\Sigma(\mathbf{x}, E)$, $\Sigma_s(\mathbf{x}, \mathbf{\Omega}' \to \mathbf{\Omega}, E' \to E)$ are nonnegative coefficients denoting the macroscopic total and the scattering cross section respectively, $\nabla_x = \left(\frac{\partial}{\partial x_1}, \frac{\partial}{\partial x_2}, \frac{\partial}{\partial x_3} \right)$ is the gradient with respect to the spatial variable $\mathbf{x}$ The boundary conditions prescribing the inflow of particles into the region D reads:

$$\varphi(\mathbf{x}, \mathbf{\Omega}, E, t) = u(\mathbf{x}, \mathbf{\Omega}, E, t), \quad (\mathbf{x}, \mathbf{\Omega}, E, t) \in \partial D_- \times]0, \tau[, \qquad (2)$$

where $\partial D_- = \{(\mathbf{x}, \mathbf{\Omega}) \in \partial D \times S^2, \ \mathbf{\Omega}.\mathbf{n}(\mathbf{x}) < 0\}$, $\mathbf{n}(\mathbf{x})$ being the outer unit normal to D at point $\mathbf{x} \in \partial D$.. The initial condition to equation (1) is given as :

$$\varphi(\mathbf{x}, \mathbf{\Omega}, E, 0) = \varphi_0(\mathbf{x}, \mathbf{\Omega}, E) \qquad (3)$$

The remainder of this chapter is structured as follows. The energy and time discretizations are presented in Section 2.2. The Section 2.3 is devoted to the discretization of the steady state single energy transport equation. The Section 2.4 deals with some iterative approaches that can be used for solving the transport equation.

2.2 Discretization of the Energy and time variables

2.2.1 Energy Discretization: Multigroup Approximation

As stated in [2], the cross-sections are continuous functions of energy E, and their dependence on E is usually complicated. There are very few instances for which simple continuous treatment dependence of cross-sections can be undertaken. Thus a discretization method has to be used. The multigroup approximation is widely used for the treatment of the energy dependency of the angular flux (see [2, 34] and the reference therein). The multigroup approximation to equation (1) begins by partitioning the energy interval into G energy groups:

$$[E_G, E_{G-1}], \cdots, [E_g, E_{g-1}], \cdots [E_1, E_0]; \quad 1 \leqslant g \leqslant G, \ E_G = 0. \tag{4}$$

The energy E_0 is chosen to be large enough for those neutrons with energy greater than E_0 to be neglected.

In the gth energy group, the angular flux is approximated by :

$$\varphi(\mathbf{x}, \mathbf{\Omega}, E, t) \approx f_g(E)\varphi_g(\mathbf{x}, \mathbf{\Omega}, t); \quad \forall E \in \ ]E_g, E_{g-1}[\,, \tag{5}$$

where the function $f_g(E)$ verifies

$$\int_{E_g}^{E_{g-1}} f_g(E)dE = 1.$$

The function φ_g is known as the gth group angular flux.

Substituting φ in (1) by its approximation (5) and integrating over $[E_g, E_{g-1}]$, gives :

$$\left[\left(\int_{E_g}^{E_{g-1}} \frac{f_g(E)}{v} dE \right) \frac{\partial}{\partial t} + \mathbf{\Omega} \cdot \nabla_x + \left(\int_{E_g}^{E_{g-1}} f_g(E)\Sigma(\mathbf{x}, E)dE \right) \right] \varphi_g(\mathbf{x}, \mathbf{\Omega}, t) =$$
$$\int_{S^2} \varphi_{g'}(\mathbf{x}, \mathbf{\Omega}', t) \left\{ \int_{E_G}^{E_0} f_{g'}(E') \left(\int_{E_g}^{E_{g-1}} \Sigma_s(\mathbf{x}, \mathbf{\Omega}, \mathbf{\Omega}', E, E')dE \right) dE' \right\} d\mathbf{\Omega}' +$$
$$\int_{E_g}^{E_{g-1}} S(\mathbf{x}, \mathbf{\Omega}, E, t)dE = \tag{6}$$
$$= \sum_{g'=1}^{G} \int_{S^2} \varphi_{g'}(\mathbf{x}, \mathbf{\Omega}', t) \left\{ \int_{E_{g'}}^{E_{g'-1}} f_{g'}(E') \left(\int_{E_g}^{E_{g-1}} \Sigma_s(\mathbf{x}, \mathbf{\Omega}' \to \mathbf{\Omega}, E' \to E)dE \right) dE' \right\} d\mathbf{\Omega}'$$
$$+ \int_{E_g}^{E_{g-1}} S(\mathbf{x}, \mathbf{\Omega}, E, t)dE,$$

which yields the following neutron transport multigroup system :

$$\left[\frac{1}{v_g} \frac{\partial}{\partial t} + \mathbf{\Omega} \cdot \nabla_x + \Sigma_{gg}(\mathbf{x}) \right] \varphi_g(\mathbf{x}, \mathbf{\Omega}, t)$$
$$= \sum_{g'=1}^{G} \int_{S^2} \Sigma_{s,gg'}(\mathbf{x}, \mathbf{\Omega}, \mathbf{\Omega}')\varphi_{g'}(\mathbf{x}, \mathbf{\Omega}', t)d\mathbf{\Omega}' + S_g(\mathbf{x}, \mathbf{\Omega}, t), \quad g = 1, \cdots, G \tag{7}$$

where :

- $\Sigma_{s,gg'}(\mathbf{x}, \mathbf{\Omega}, \mathbf{\Omega'}) = \int_{E_{g'}}^{E_{g'-1}} f_{g'}(E') \left(\int_{E_g}^{E_{g-1}} \Sigma_s(\mathbf{x}, \mathbf{\Omega'} \to \mathbf{\Omega}, E' \to E) dE \right) dE'$: is the inter-group scattering cross-section ;

- $\Sigma_{gg}(\mathbf{x}) = \int_{E_g}^{E_{g-1}} f_g(E) \sigma(\mathbf{x}, E) dE$: is the gth group total cross-section ;

- $\frac{1}{v_g} = \int\limits_{E_g}^{E_{g-1}} \frac{f_g(E)}{v} dE$: is the gth group harmonic mean speed ;

- $S_g(\mathbf{x}, \mathbf{\Omega}, t) = \int_{E_g}^{E_{g-1}} S(\mathbf{x}, \mathbf{\Omega}, E, t) dE$: is the gth group source function.

The group initial condition is obtained from the initial condition (2) and the approximation (5) as follows

$$\varphi_g(\mathbf{x}, \mathbf{\Omega}, 0) = \varphi_{g0}(\mathbf{x}, \mathbf{\Omega}) \tag{8}$$

where $\varphi_{g0}(\mathbf{x}, \mathbf{\Omega}) = f_g(E) \varphi_{g0}(\mathbf{x}, \mathbf{\Omega}, E)$.

Similarly, for the boundary conditions (3), one has

$$\varphi_g(\mathbf{x}, \mathbf{\Omega}, t) = u_g(\mathbf{x}, \mathbf{\Omega}, t), \quad \forall (\mathbf{x}, \mathbf{\Omega}, t) \in \partial D_- \times]0, \tau[, \tag{9}$$

where $u_g(\mathbf{x}, \mathbf{\Omega}, t) = f_g(E)(\mathbf{x}, \mathbf{\Omega}, E, t)$.

Setting

$$L_{gg}\varphi_g = [\mathbf{\Omega} \cdot \nabla_x + \Sigma_{gg}] \varphi_g \tag{10}$$

and

$$K_{gg'}\varphi_g = \int_{S^2} \Sigma_{s,gg'}(\mathbf{x}, \mathbf{\Omega}, \mathbf{\Omega'}) \varphi_{g'}(\mathbf{x}, \mathbf{\Omega'}, t) d\mathbf{\Omega'}, \tag{11}$$

the multigroup neutron transport operator $\mathbf{T} = (T_{gg'})_{1 \leq g, g' \leq G}$ is defined by (see [29]) :

$$T_{gg'}\varphi_g = \delta_{gg'} L_{gg} - K_{gg'}, \quad 1 \leq g, g' \leq G, \tag{12}$$

where $\delta_{gg'}$ is the Kronecker delta. Therefore, the time dependent multigroup equation (7) with the associated boundary condition (9) and the initial condition (8) can be written as follows

$$\frac{1}{v_g} \frac{\partial \mathbf{\Psi}}{\partial t} + \mathbf{T}\mathbf{\Psi} = \mathbf{S}, \tag{13}$$

given that

$$\mathbf{\Psi}(\mathbf{x}, \mathbf{\Omega}, 0) = \mathbf{\Psi}_0(\mathbf{x}, \mathbf{\Omega}), \tag{14}$$

where $\mathbf{\Psi} = \begin{pmatrix} \varphi_1 \\ \vdots \\ \varphi_G \end{pmatrix}$, $\mathbf{\Psi}_0 = \begin{pmatrix} \varphi_{1,0} \\ \vdots \\ \varphi_{G,0} \end{pmatrix}$ and $\mathbf{S} = \begin{pmatrix} S_1 \\ \vdots \\ S_G \end{pmatrix}$.

Here, the multigroup operator $\mathbf{T}$ absorbs the boundary conditions.

Assuming that all scattering leads to a loss of kinetic energy by a neutron, the inter-group scattering cross-sections have the properties (see [2, 29]) :

$$\begin{cases} \Sigma_{s,gg'} > 0 \ \ if \ \ g' \le g \\ \Sigma_{s,gg'} = 0 \ \ if \ \ g' > g. \end{cases} \tag{15}$$

Therefore, the matrix of operator $\mathbf{T}$ is lower triangular and and the time-dependent group flux may be obtained systematically starting from the highest energy group $g = 1$. The typical group (monoenergetic or one-group) equation is written as :

$$\begin{aligned} &\frac{1}{v_g}\frac{\partial}{\partial t}\varphi_g(\mathbf{x},\mathbf{\Omega},t) + \mathbf{\Omega}\cdot\nabla_x\varphi_g(\mathbf{x},\mathbf{\Omega},t) + \Sigma_{gg}(\mathbf{x})\varphi_g(\mathbf{x},\mathbf{\Omega},t) \\ &- \int_{S^2}\Sigma_{s,gg}(\mathbf{x},\mathbf{\Omega},\mathbf{\Omega}')\varphi_g(\mathbf{x},\mathbf{\Omega}',t)d\mathbf{\Omega}' = \mathbf{S}_g(\mathbf{x},\mathbf{\Omega},t) \quad (g = 1,\cdots,G.) \end{aligned} \tag{16}$$

where

$$\mathbf{S}_g(\mathbf{x},\mathbf{\Omega},t) = S_g(\mathbf{x},\mathbf{\Omega},t) + \sum_{g'=1}^{g-1}\int_{S^2}\Sigma_{s,gg'}(\mathbf{x},\mathbf{\Omega},\mathbf{\Omega}')\varphi_{g'}(\mathbf{x},\mathbf{\Omega}',t)d\mathbf{\Omega}' \quad (g = 1,\cdots,G.) \tag{17}$$

The sweep through all energy group $g = 1,\cdots,G$ is referred to as an outer iteration. For the general case where upscatter is included, the multigroup equation can be solved by using iterative methods such as Jacobi-like and Gauss-Seidel like iterations. The Gauss-Seidel like iteration for the multigroup equation (13) reads : Given an initial guest $\mathbf{\Psi}^{(0)}$, for $l = 0,1,\cdots$ until $\{\mathbf{\Psi}^{(l)}\}$ converges, solve

$$\left[\frac{1}{v_g}\frac{\partial}{\partial t} + \mathbf{T}_{gg}\right]\varphi_g^{(l+1)} = S_g - \sum_{g'<g}T_{gg'}\varphi_g^{(l+1)} - \sum_{g'>g}T_{gg'}\varphi_g^{(l)}, \ \ g = 1,\cdots,G. \tag{18}$$

By removing the temporal dependence of the multigroup flux $\mathbf{\Psi}$ and assuming that the multigroup source function $\mathbf{S}$ and the boundary conditions are independent of time t, the time independent or the steady state form of the multigroup neutron transport problem (13) is obtained and reads:

$$\mathbf{T}\mathbf{\Psi} = \mathbf{S}, \tag{19}$$

Therefore, under the assumption (15), the steady state typical group equation is obtained as follows :

$$\begin{aligned} &\mathbf{\Omega}\cdot\nabla_x\varphi_g(\mathbf{x},\mathbf{\Omega}) + \Sigma_{gg}(\mathbf{x})\varphi_g(\mathbf{x},\mathbf{\Omega}) - \int_{S^2}\Sigma_{s,gg}(\mathbf{x},\mathbf{\Omega},\mathbf{\Omega}')\varphi_g(\mathbf{x},\mathbf{\Omega}')d\mathbf{\Omega}' = \mathbf{S}_g(\mathbf{x},\mathbf{\Omega}) \\ &(g = 1,\cdots,G.) \end{aligned} \tag{20}$$

where

$$\mathbf{S}_g(\mathbf{x}, \mathbf{\Omega}) = S_g(\mathbf{x}, \mathbf{\Omega}) + \sum_{g'=1}^{g-1} \int_{S^2} \Sigma_{s,gg'}(\mathbf{x}, \mathbf{\Omega}, \mathbf{\Omega}')\varphi_{g'}(\mathbf{x}, \mathbf{\Omega}')d\mathbf{\Omega}' \quad (g = 1, \cdots, G.) \qquad (21)$$

The Numerical solutions for the energy-dependent neutron transport are based on numerical solution for the typical group equation.

2.2.2 Time Discretization

We consider in this subsection the gth $(1 \le g \le G)$ group time-dependent equation written as :

$$\begin{aligned} &\tfrac{1}{v_g}\tfrac{\partial}{\partial t}\varphi_g(\mathbf{x}, \mathbf{\Omega}, t) + \mathbf{\Omega} \cdot \nabla_x \varphi_g(\mathbf{x}, \mathbf{\Omega}, t) + \Sigma_{gg}(\mathbf{x})\varphi_g(\mathbf{x}, \mathbf{\Omega}, t) \\ &- \int_{S^2} \Sigma_{s,gg}(\mathbf{x}, \mathbf{\Omega}, \mathbf{\Omega}')\varphi_g(\mathbf{x}, \mathbf{\Omega}', t)d\mathbf{\Omega}' = \mathbf{S}_g(\mathbf{x}, \mathbf{\Omega}, t) \end{aligned} \qquad (22)$$

with the associated boundary condition (9) and the initial condition (8).

The treatment of the time-dependent transport equation (22) can be done by using varieties of methods such as finite difference and variational methods. Finite-differencing the derivative in time is the widely used approach. We briefly present here an implicit time discretization of equation (22).

The time interval $]0, \tau[$ is divided into N time steps as follows:

$$t_0 = 0, \quad t_N = \tau, \quad t_{n+1} = t_n + \Delta t \ (0 \le n < N)$$

where $\Delta t = \frac{\tau}{N}$ is the time step. By integrating equation (22) from $t = n$ to $t = n+1$ and dividing by Δt, we obtain :

$$\begin{aligned} &\tfrac{1}{v_g \Delta t}[\varphi_{g,n+1}(\mathbf{x}, \mathbf{\Omega}) - \varphi_{g,n}(\mathbf{x}, \mathbf{\Omega})] + \mathbf{\Omega} \cdot \nabla_x \bar{\varphi}_g(\mathbf{x}, \mathbf{\Omega}) + \Sigma_{gg}(\mathbf{x})\bar{\varphi}_g(\mathbf{x}, \mathbf{\Omega}) \\ &- \int_{S^2} \Sigma_{s,gg}(\mathbf{x}, \mathbf{\Omega}, \mathbf{\Omega}')\bar{\varphi}_g(\mathbf{x}, \mathbf{\Omega}')d\mathbf{\Omega}' = \bar{\mathbf{S}}_g(\mathbf{x}, \mathbf{\Omega}) \end{aligned} \qquad (23)$$

where $\varphi_{g,i}(\mathbf{x}, \mathbf{\Omega}) = \varphi_{g,i}(\mathbf{x}, \mathbf{\Omega}, t_i)$, $\bar{\varphi}_g(\mathbf{x}, \mathbf{\Omega})$ and $\bar{\mathbf{S}}_g(\mathbf{x}, \mathbf{\Omega})$ are the average flux and source respectively defined by :

$$\bar{\varphi}_g(\mathbf{x}, \mathbf{\Omega}) = \frac{1}{\Delta t} \int_{t_n}^{t_{n+1}} \varphi_g(\mathbf{x}, \mathbf{\Omega}, t)dt \qquad (24)$$

$$\bar{\mathbf{S}}_g(\mathbf{x}, \mathbf{\Omega}) = \frac{1}{\Delta t} \int_{t_n}^{t_{n+1}} \mathbf{S}_g(\mathbf{x}, \mathbf{\Omega}, t)dt \qquad (25)$$

The angular flux $\varphi_{g,n}(\mathbf{x}, \mathbf{\Omega})$ is known from the previous calculation at time t_n and it serves as initial condition for the calculation at the time t_{n+1}. There remains two unknown quantities : $\varphi_{g,n+1}(\mathbf{x}, \mathbf{\Omega})$ and $\bar{\varphi}_g(\mathbf{x}, \mathbf{\Omega})$, for the equation (23). There is need of one supplementary relation between $\varphi_{g,n+1}(\mathbf{x}, \mathbf{\Omega})$ and $\bar{\varphi}_g(\mathbf{x}, \mathbf{\Omega})$. The common choice is to assume that :

$$\bar{\varphi}_g(\mathbf{x}, \mathbf{\Omega}) = \beta\varphi_{g,n+1}(\mathbf{x}, \mathbf{\Omega}) + (1 - \beta)\varphi_{g,n}(\mathbf{x}, \mathbf{\Omega}) \tag{26}$$

for an appropriate value for β. The well known choices for β are (see [2]) :

- $\beta = \frac{1}{2}$: the unconditionally stable Crank-Nicholson scheme;

- $\beta = \frac{2}{3}$: the Galerkin scheme;

- $\beta = 1$: the backward Euler scheme.

Using equation (26) to eliminate $\varphi_{g,n+1}(\mathbf{x}, \mathbf{\Omega})$ in equation (23) yields :

$$\mathbf{\Omega} \cdot \nabla_x \bar{\varphi}_g(\mathbf{x}, \mathbf{\Omega}) + \left(\Sigma_{gg}(\mathbf{x}) + \frac{1}{\beta v_g \Delta t}\right) \bar{\varphi}_g(\mathbf{x}, \mathbf{\Omega}) - \int_{S^2} \Sigma_{s,gg}(\mathbf{x}, \mathbf{\Omega}, \mathbf{\Omega}')\bar{\varphi}_g(\mathbf{x}, \mathbf{\Omega}')d\mathbf{\Omega}'$$
$$= \left(\bar{\mathbf{S}}_g(\mathbf{x}, \mathbf{\Omega}) + \frac{\varphi_{g,n}(\mathbf{x}, \mathbf{\Omega})}{\beta v_g \Delta t}\right) \tag{27}$$

Defining the effective total cross section as :

$$\tilde{\Sigma}_{gg}(\mathbf{x}) = \left(\Sigma_{gg}(\mathbf{x}) + \frac{1}{\beta v_g \Delta t}\right) \tag{28}$$

and the effective source at time t_{n+1} as :

$$\tilde{\mathbf{S}}_{g,n+1}(\mathbf{x}, \mathbf{\Omega}) = \left(\bar{\mathbf{S}}_g(\mathbf{x}, \mathbf{\Omega}) + \frac{\varphi_{g,n}(\mathbf{x}, \mathbf{\Omega})}{\beta v_g \Delta t}\right), \tag{29}$$

the equation (27) reads

$$\mathbf{\Omega} \cdot \nabla_x \bar{\varphi}_g(\mathbf{x}, \mathbf{\Omega}) + \tilde{\Sigma}_{gg}(\mathbf{x})\bar{\varphi}_g(\mathbf{x}, \mathbf{\Omega}) - \int_{S^2} \Sigma_{s,gg}(\mathbf{x}, \mathbf{\Omega}, \mathbf{\Omega}')\bar{\varphi}_g(\mathbf{x}, \mathbf{\Omega}')d\mathbf{\Omega}' = \tilde{\mathbf{S}}_{g,n+1}(\mathbf{x}, \mathbf{\Omega}). \tag{30}$$

The boundary condition (9) at time t_{n+1} is given by :

$$\bar{\varphi}_g(\mathbf{x}, \mathbf{\Omega}) = u_g(\mathbf{x}, \mathbf{\Omega}, t_{n+1}) = u_{g,n+1}(\mathbf{x}, \mathbf{\Omega}), \quad \forall \mathbf{x} \in \partial D_-. \tag{31}$$

The problem (30)-(31) is in the same form as the gth group stationary neutron transport problem.

The solution of time-dependent multi-group problems involves solving many problems of the form (30)-(31). Thus the development of fast solution method for the time-dependent multi-group problems requires accurate and efficient solution methods for the monoenergetic steady state problem [11].

2.3 Numerical Methods for the Steady State Monoenergetic Problem

We consider in this section the general single group, steady state first order neutron transport equation. Let $D \subset \mathbb{R}^3$ be the physical domain and $\varphi(\mathbf{x}, \mathbf{\Omega})$ be the flux of neutrons at position $\mathbf{x} \in D$ travelling in the direction $\mathbf{\Omega} \in S^2$. Then

$$\mathbf{\Omega} \cdot \nabla_x \varphi(\mathbf{x}, \mathbf{\Omega}) + \Sigma(\mathbf{x})\varphi(\mathbf{x}, \mathbf{\Omega}) = K\varphi(\mathbf{x}, \mathbf{\Omega}) + q(\mathbf{x}, \mathbf{\Omega}), \quad (\mathbf{x}, \mathbf{\Omega}) \in D \times S^2 \quad (32)$$

$$\varphi(\mathbf{x}, \mathbf{\Omega}) = u(\mathbf{x}, \mathbf{\Omega}), \quad \mathbf{x} \in \partial D_-, \quad (33)$$

where the integral operator K is defined by

$$K\varphi(\mathbf{x}, \mathbf{\Omega}) = \int_{S^2} \Sigma_s(\mathbf{x}, \mathbf{\Omega}, \mathbf{\Omega}')\varphi(\mathbf{x}, \mathbf{\Omega})d\mathbf{\Omega}'. \quad (34)$$

It is commonly assumed that the physical domain on which the single-group equation is posed consists of only isotropic media (see [2, 11, 49, 47]). In this case, the kernel Σ_s of the collision operator K depends on $\mathbf{\Omega}' \cdot \mathbf{\Omega}$, which is equal to the cosine of the scattering angle θ_0 (see **Figure 4**), since $\mathbf{\Omega}'$ and $\mathbf{\Omega}$ have unit length. Thus

$$\Sigma_s(\mathbf{x}, \mathbf{\Omega}, \mathbf{\Omega}') = \Sigma_s(\mathbf{x}, \mathbf{\Omega} \cdot \mathbf{\Omega}') = \Sigma_s(\mathbf{x}, \mu_0), \quad (35)$$

where $\mu_0 = \cos(\theta_0)$. Therefore, the scattering kernel can be expanded in terms of orthogonal Legendre polynomials as:

$$\Sigma_s(\mathbf{x}, \mu_0) = \sum_{l=0}^{\infty} \frac{2l+1}{4\pi} \Sigma_{s,l}(\mathbf{x}) P_l(\mu_0). \quad (36)$$

where the Legendre polynomials are defined by the Rodrigues' formula as,

$$P_l(\mu) = \frac{1}{2^l l!} \frac{d^l}{d\mu^l} \left[(\mu^2 - 1)^l \right]. \quad (37)$$

$$\Sigma_{s,l}(\mathbf{x}) = 2\pi \int_{-1}^{1} \Sigma_s(\mathbf{x}, \mu_0) P_l(\mu_0) d\mu_0. \quad (38)$$

Let $Y_{lm}(\mathbf{\Omega}) = Y_{lm}(\theta, \phi)$ denote the the spherical harmonics functions defined as:

$$Y_{lm}(\theta, \phi) = \sqrt{\frac{(2l+1)(l-m)!}{4\pi(l+m)!}} P_l^m(\cos\theta)e^{im\phi}, \quad (39)$$

where

$$P_l^m(\mu) = (-1)^m (1-\mu^2)^{\frac{m}{2}} \frac{d^m P_l(\mu)}{d\mu^m} \quad (40)$$

34

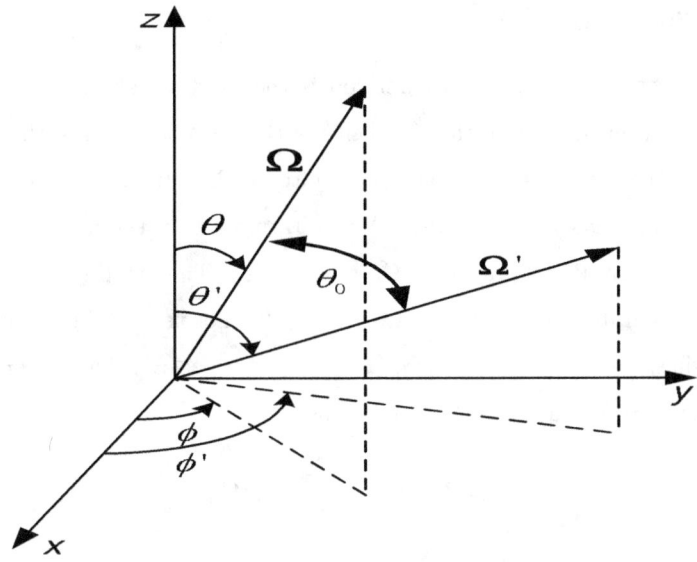

Figure 4: Representation of the scattering angle θ_0

are the associated Legendre polynomials.

If θ_0 is the angle between directions $\mathbf{\Omega} = (\theta, \Phi)$ and $\mathbf{\Omega}' = (\theta', \Phi')$, it holds that:

$$P_l(\mu_0) \;=\; \frac{4\pi}{2l+1} \sum_{m=-l}^{l} Y_{lm}(\mathbf{\Omega}) Y_{lm}^*(\mathbf{\Omega}') \tag{41}$$

$$=\; P_l(\mu) P_l(\mu') + 2 \sum_{k=1}^{l} P_l^k(\mu) P_l^k(\mu') \frac{(l-k)!}{(l+k)!} \cos k(\Phi - \Phi'), \tag{42}$$

where $\mu_0 = \cos\theta_0$, $\mu = \cos\theta$ and $\mu' = \cos\theta'$.

The single-group equation (32) has two types of independent variables: the angular and the spatial variables. The numerical solution of the problem (32)-(33) requires discretization of both angular and spatial variables. Spatial and angular discretizations are generally performed independently.

2.3.1 Angular Discretization

The angular discretization is generally accomplished in two ways: the discrete ordinates method (S_N) and the expansion of the angular flux in terms of angular basis functions. The spherical harmonics method (P_N) is a particular case of the second approach.

35

Discrete Ordinates Method

The discrete ordinate (S_N) approximation is the most widely used method for the angular discretization because of the simplicity of the formulation. For this approach, it is assumed that the neutrons travel only in a finite number M of specified directions $\mathbf{\Omega}_n$ ($1 \leq n \leq M$), called *discrete ordinates*. A quadrature set on the unit sphere is formed by associating to each discrete ordinate $\mathbf{\Omega}_n$ a weight ω_n. The quadrature values $(\mathbf{\Omega}_n, \omega_n)$ are selected to accurately evaluate the integral in (32). Let us introduce a finite set of M quadrature values $\{(\mathbf{\Omega}_n, \omega_n)\}_{n=1}^{n=M}$, where $\mathbf{\Omega}_n = (\mu_n, \eta_n, \xi_n) \in S^2$. The integral expression over the unit sphere S^2 may be approximated as follows:

$$\int_{S^2} f(\mathbf{\Omega}) d\mathbf{\Omega} \approx \sum_{n=1}^{M} \omega_n f(\mathbf{\Omega}_n). \tag{43}$$

For the selection of the quadrature set, it is generally required that (see [53]) :

- The associated weights must be positive and verify

$$\sum_{i=1}^{M} \omega_i = 4\pi. \tag{44}$$

- The quadrature set is chosen to be symmetric over the unit sphere, so the solution will be invariant with respect to a 90 degree rotation and reflection. It follows that

$$\sum_{i=1}^{M} \omega_i \mu_i^n = \sum_{i=1}^{M} \omega_i \xi_i^n = \sum_{i=1}^{K} \omega_i \eta_i^n = 0, \quad \text{for } n \text{ odd.} \tag{45}$$

- The quadrature set should lead to accurate values for moments of the angular flux. That is:

$$\sum_{i=1}^{M} \omega_i \mu_i^n = \sum_{i=1}^{M} \omega_i \xi_i^n = \sum_{i=1}^{K} \omega_i \eta_i^n = \frac{4\pi}{n+1}, \quad \text{for } n \text{ even.} \tag{46}$$

Using the quadrature formula (43) for the approximation of the integral term in equation (32), we obtain the following angularly discretised formulation of the equation (32) :

$$\mathbf{A}_1 \frac{\partial \Phi}{\partial x}(\mathbf{x}) + \mathbf{A}_2 \frac{\partial \Phi}{\partial y}(\mathbf{x}) + \mathbf{A}_3 \frac{\partial \Phi}{\partial z}(\mathbf{x}) + \mathbf{\Sigma}_1 \Phi(\mathbf{x}) - \mathbf{K}_1 \Phi(\mathbf{x}) = \mathbf{Q}_1(\mathbf{x}), \tag{47}$$

where

- $\Phi(\mathbf{x}) = \begin{pmatrix} \varphi_1(\mathbf{x}) \\ \cdot \\ \varphi_M(\mathbf{x}) \end{pmatrix}$ with $\varphi_l(\mathbf{x}) = \varphi(\mathbf{x}, \mathbf{\Omega}_j)$, $1 \leq j \leq M$.

36

- $\mathbf{Q}(\mathbf{x}) = \begin{pmatrix} q_1(\mathbf{x}) \\ \cdot \\ q_M(\mathbf{x}) \end{pmatrix}$ with $q_l(\mathbf{x}) = q(\mathbf{x}, \mathbf{\Omega}_j)$, $1 \le j \le M$.

- $\mathbf{A}_1$, $\mathbf{A}_1$, $\mathbf{A}_3$ and $\mathbf{\Sigma}_1$ are diagonal matrices defined as

$$\begin{aligned}
\mathbf{A}_1 = diag(\mu_1, \cdots, \mu_M) \quad &\mathbf{A}_2 = diag(\eta_1, \cdots, \eta_M) \\
\mathbf{A}_3 = diag(\xi_1, \cdots, \xi_M) \quad &\mathbf{\Sigma}_1 = diag(\Sigma(\mathbf{x}), \cdots, \Sigma(\mathbf{x}))
\end{aligned} \tag{48}$$

- $\mathbf{K}_1$ is the $M \times M$ discrete ordinates scattering operator given by :

$$\mathbf{K}_{1,ij} = \omega_j \Sigma_s(\mathbf{x}, \mathbf{\Omega}_i, \mathbf{\Omega}'_j). \tag{49}$$

Expansion in terms of angular basis functions

Another approach to obtain the angular discretised form of the transport equation (32) consists of expanding the angular variable and generate the weak form equations through weighted residual method (see [26]). The angular flux φ is approximated by the expansion

$$\varphi(\mathbf{x}, \mathbf{\Omega}) \approx \sum_{j=1}^{M} \phi_j(\mathbf{x}) h_j(\mathbf{\Omega}), \tag{50}$$

where $h_m(\mathbf{\Omega})$ $(j = 1, \cdots, M)$ are the angular basis functions and $\phi_j(\mathbf{x})$ are their corresponding angular moments. Using the Galerkin scheme by inserting the approximation (50) into the transport equation (32), multiplying by the weighting function $h_m(\mathbf{\Omega})$ $(j = 1, \cdots, M)$ and integrating the resulting equation over the angular variable, the following form of the angularly discretised transport equation is obtained (see [26]) :

$$\mathbf{A}_x \frac{\partial \Phi}{\partial x}(\mathbf{x}) + \mathbf{A}_y \frac{\partial \Phi}{\partial y}(\mathbf{x}) + \mathbf{A}_z \frac{\partial \Phi}{\partial z}(\mathbf{x}) + \mathbf{\Sigma} \Phi(\mathbf{x}) - \mathbf{K} \Phi(\mathbf{x}) = \mathbf{Q}(\mathbf{x}), \tag{51}$$

where

- $\Phi(\mathbf{x})$ is the vector of M angular moments that give the approximation of the angular flux at position $\mathbf{x}$;

- $\mathbf{Q}(\mathbf{x})$ is the vector of size M containing the angularly discretised source at $\mathbf{x}$:

$$\mathbf{Q}_i(\mathbf{x}) = \int_{S^2} q(\mathbf{x}, \mathbf{\Omega}) h_i(\mathbf{\Omega}) d\mathbf{\Omega}; \tag{52}$$

- $\mathbf{A}_x$, $\mathbf{A}_y$, $\mathbf{A}_z$ and $\boldsymbol{\Sigma}$ are $M \times M$ matrices defined by :

$$\mathbf{A}_{x,ij} = \int_{S^2} h_i(\boldsymbol{\Omega})h_j(\boldsymbol{\Omega})\Omega_x d\Omega \tag{53}$$

$$\mathbf{A}_{y,ij} = \int_{S^2} h_i(\boldsymbol{\Omega})h_j(\boldsymbol{\Omega})\Omega_y d\Omega \tag{54}$$

$$\mathbf{A}_{z,ij} = \int_{S^2} h_i(\boldsymbol{\Omega})h_j(\boldsymbol{\Omega})\Omega_z d\Omega \tag{55}$$

$$\boldsymbol{\Sigma}_{ij} = \int_{S^2} \Sigma(\mathbf{x})h_i(\boldsymbol{\Omega})h_j(\boldsymbol{\Omega})d\Omega \tag{56}$$

- $\mathbf{K}$ is the $M \times M$ scattering operator given by :

$$\mathbf{K}_{ij} = \int_{S^2} h_i(\boldsymbol{\Omega}) \left[\int_{S^2} \Sigma_s(\mathbf{x}, \boldsymbol{\Omega}, \boldsymbol{\Omega}')h_j(\boldsymbol{\Omega}')d\Omega' \right] d\Omega. \tag{57}$$

In the isotropic media, by using the expansion (36) of the kernel Σ_s of the collision operator, the coefficients of the scattering matrix $\mathbf{K}$ are given by ([27] for the full details derivation) :

$$\mathbf{K}_{ij} = \sum_{l=0}^{\infty} \Sigma_{s,l}(\mathbf{x})\alpha_i^{e,l,0}\alpha_j^{e,l,0} + 2\sum_{l=1}^{\infty} \Sigma_{sl}(\mathbf{x}) \sum_{m=1}^{l} \left[\alpha_i^{e,l,m}\alpha_j^{e,l,m} + \alpha_i^{o,l,m}\alpha_j^{o,l,m} \right]. \tag{58}$$

The coefficients α_i are given by :

$$\alpha_i^{e,l,m} = \int_{S^2} h_i(\boldsymbol{\Omega})Y_{lm}^e(\boldsymbol{\Omega})d\Omega \tag{59}$$

$$\alpha_i^{o,l,m} = \int_{S^2} h_i(\boldsymbol{\Omega})Y_{lm}^o(\boldsymbol{\Omega})d\Omega \tag{60}$$

where $Y_{lm}^e(\boldsymbol{\Omega})$ and $Y_{lm}^o(\boldsymbol{\Omega})$ are the real and complex parts of the spherical harmonic function $Y_{lm}(\boldsymbol{\Omega})$:

$$Y_{lm}^e(\boldsymbol{\Omega}) = Y_{lm}(\theta, \phi) = \sqrt{\frac{(2l+1)(l-m)!}{4\pi(l+m)!}} \, P_l^m(\cos\theta) \cos m\phi, \tag{61}$$

$$Y_{lm}^o(\boldsymbol{\Omega}) = Y_{lm}(\theta, \phi) = \sqrt{\frac{(2l+1)(l-m)!}{4\pi(l+m)!}} \, P_l^m(\cos\theta) \sin m\phi, \tag{62}$$

There are many ways for choosing the basis functions $h_i(\boldsymbol{\Omega})$. If they are the spherical harmonics functions through order N, we obtained the *spherical harmonics* or P_N method. The finite element method in angle can also be used for the construction of the angular basis functions $h_i(\boldsymbol{\Omega})$. Other class of functions such as wavelets functions (see [26, 27, 28]) and Walsh functions (see [42]) have been used angular expansion of the flux.

2.3.2 Spatial Discretization

For the spatial discretization of the angularly discretised form (47) or (51) of the single group equation (32), a spatial grid is imposed on the physical system. Then, each equation of the system (47) or (51) is discretised on this grid. Several spatial discretization techniques for the transport operator have been proposed. Among them are **finite difference**, **finite element** and **nodal** methods (see [2, 3, 4, 18, 19, 49, 48, 29, 34, 23, 61, 60]).

Finite Differences Schemes

Let us consider the following numerical grid of the spatial domain :

$$D_h = \{(x_i, y_j, z_k), 0 \le i \le I, 0 \le i \le J, 0 \le i \le K\}, \tag{63}$$

where $x_i = x_{i-1} + \Delta x_i$, $y_j = y_{j-1} + \Delta y_j$, $z_j = z_{k-1} + \Delta z_k$. The cell center grid points are defined as: $x_{i+\frac{1}{2}} = \dfrac{x_{i+1} - x_i}{2}$, $y_{j+\frac{1}{2}} = \dfrac{y_{j+1} - y_j}{2}$, $z_{k+\frac{1}{2}} = \dfrac{z_{k+1} - z_k}{2}$. The cross-sections are assumed to be constant in each cell and denoted by $\Sigma_{i+\frac{1}{2}j+\frac{1}{2}k+\frac{1}{2}}$. By integrating the equation (47) over the cell $V_{ijk} = [x_{i-1}, x_i] \times [y_{j-1}, x_j] \times [z_{k-1}, z_k]$ and the and dividing by the cell volume $\Delta x_i \Delta y_j \Delta z_k$, we obtain the following equation for the discrete direction Ω_l :

$$\frac{\mu_l}{\Delta x_i} \left(\varphi_{l,i,j+\frac{1}{2},k+\frac{1}{2}} - \varphi_{l,i-1,j+\frac{1}{2},k+\frac{1}{2}} \right) + \frac{\eta_l}{\Delta y_j} \left(\varphi_{l,i+\frac{1}{2},j,k+\frac{1}{2}} - \varphi_{l,i+\frac{1}{2},j-1,k+\frac{1}{2}} \right) +$$

$$\frac{\xi_l}{\Delta z_k} \left(\varphi_{l,i+\frac{1}{2},j+\frac{1}{2},k} - \varphi_{l,i+\frac{1}{2},j+\frac{1}{2},k-1} \right) + \Sigma_{i+\frac{1}{2}j+\frac{1}{2}k+\frac{1}{2}} \varphi_{l,i+\frac{1}{2},j+\frac{1}{2},k+\frac{1}{2}} = Q_{l,i+\frac{1}{2},j+\frac{1}{2},k+\frac{1}{2}} \tag{64}$$

where

$$\varphi_{l,i,j+\frac{1}{2},k+\frac{1}{2}} = \frac{1}{\Delta y_j \Delta z_k} \int_{y_{j-1}}^{y_j} \int_{z_{k-1}}^{z_k} \varphi_l(x_i, y, z) dy dz;$$

$$\varphi_{l,i+\frac{1}{2},j,k+\frac{1}{2}} = \frac{1}{\Delta x_i \Delta z_k} \int_{x_{i-1}}^{x_i} \int_{z_{k-1}}^{z_k} \varphi_l(x, y_j, z) dx dz;$$

$$\varphi_{l,i+\frac{1}{2},j+\frac{1}{2},k} = \frac{1}{\Delta x_i \Delta y_j} \int_{x_{i-1}}^{x_i} \int_{y_{j-1}}^{y_j} \varphi_l(x, y, z_k) dx dy;$$

define the fluxes average over the cell faces and

$$\varphi_{l,i+\frac{1}{2},j+\frac{1}{2},k+\frac{1}{2}} = \frac{1}{\Delta x_i \Delta y_j \Delta z_k} \int_{x_{i-1}}^{x_i} \int_{y_{j-1}}^{y_j} \int_{z_{k-1}}^{z_k} \varphi_l(\mathbf{x}) dx dy dz;$$

$$Q_{l,i+\frac{1}{2},j+\frac{1}{2},k+\frac{1}{2}} = \frac{1}{\Delta x_i \Delta y_j \Delta z_k} \int_{x_{i-1}}^{x_i} \int_{y_{j-1}}^{y_j} \int_{z_{k-1}}^{z_k} \left[\sum_{l'=1}^{M} \mathbf{K}_{1,ll'} \varphi_{l'}(\mathbf{x}) - q_l(\mathbf{x}) \right] dx dy dz.$$

are the cell center values. The equation (64) is exact in the sense that it contains no spatial approximation. A similar discrete form equation can be obtained from equation. (51).

Weighted Differences Schemes

For the approximation of the cell average flux, the following weighted (WD) difference schemes (see [53]) can be used :

$$\varphi_{l,i+\frac{1}{2},j+\frac{1}{2},k+\frac{1}{2}} = a_{l,ijk}\varphi_{l,i,j+\frac{1}{2},k+\frac{1}{2}} + (1 - a_{l,ijk})\varphi_{l,i-1,j+\frac{1}{2},k+\frac{1}{2}} \tag{65}$$

$$\varphi_{l,i+\frac{1}{2},j+\frac{1}{2},k+\frac{1}{2}} = b_{l,ijk}\varphi_{l,i+\frac{1}{2},j,k+\frac{1}{2}} + (1 - b_{l,ijk})\varphi_{l,i+\frac{1}{2},j-1,k+\frac{1}{2}} \tag{66}$$

$$\varphi_{l,i+\frac{1}{2},j+\frac{1}{2},k+\frac{1}{2}} = c_{l,ijk}\varphi_{l,i+\frac{1}{2},j+\frac{1}{2},k} + (1 - c_{l,ijk})\varphi_{l,i+\frac{1}{2},j+\frac{1}{2},k-1} \tag{67}$$

where the values $a_{l,ijk}, b_{l,ijk}$ and $c_{l,ijk}$ are determined based on the type of the weighted scheme employed:

- For the diamond difference scheme (DSN) [34, 53] the weights are set to constant values: $a_{l,ijk} = b_{l,ijk} = c_{l,ijk} = \frac{1}{2}$. The DSN scheme generates physically meaningless negative fluxes in some circumstances (in regions where flux gradient is large). In this case, the *negative flux fix up* strategy where the negative fluxes are set to zero and the cell average angular flux is recalculated to preserve the balance of particles, is utilized.

- For the directional theta-weighted scheme (DTW), the weights are restricted to range between 0.5 and 1.0. for each spatial node and direction, The weights are chosen such that the positivity of the angular flux is ensured.

- The exponential directional weight scheme (EDW) [53] uses the DTW to predict a solution that is corrected using an exponential fit. The EDW is an inherently positive scheme.

The following discrete equations are obtained when the first order upwind discretization is used:

$$\frac{\mu_l}{\Delta x_i}\left(\varphi_{l,i,j,k} - \varphi_{l,i-1,j,k}\right) + \frac{\eta_l}{\Delta y_j}\left(\varphi_{l,i,j,k} - \varphi_{l,i,j-1,k}\right) +$$

$$\frac{\xi_l}{\Delta z_k}\left(\varphi_{l,i,j,k} - \varphi_{l,i,j,k-1}\right) + \Sigma_{ijk}\varphi_{l,i,j,k} = Q_{l,i,j,k} \tag{68}$$

where $f_{l,i,j,k} \approx f_l(x_i, y_j, z_k)$ and

$$Q_l(\mathbf{x}) = \sum_{l'=1}^{M} \mathbf{K}_{1,ll'}\varphi_{l'}(\mathbf{x}) + q_l(\mathbf{x}).$$

Step Characteristic Schemes (SC)

In the method of characteristic, the transport equation is solved analytically along the straight lines at the discrete number of directions within a computational cell. This lines known as characteristic directions are regarded as the neutron paths along which the differential operator of the Boltzmann equation reduces to a total differential. The outgoing angular flux is solved along the s-axis, where this axis is oriented along the characteristic direction $\mathbf{\Omega}$ (see [35]). Let us l^{th} equation of system (47) stated as follows:

$$\mathbf{\Omega}_l \cdot \nabla_x \varphi_l(\mathbf{x}) + \Sigma(\mathbf{x})\varphi_l(\mathbf{x}) = Q_l(x). \tag{69}$$

The streaming term can be written as:

$$\mathbf{\Omega}_l \cdot \nabla_x \varphi_l(\mathbf{x}) = \frac{d\varphi_l(s)}{ds}. \tag{70}$$

Setting $Q_l(s) = Q_l(x)$ and $\Sigma(x) = \Sigma(\mathbf{x})$, we obtain the following characteristic form of the equation (69)

$$\frac{d\varphi_l}{ds} + \Sigma\varphi_l = Q_l \tag{71}$$

which has a solution of the form [35]

$$\varphi_l(s) = \varphi_{l0} \exp(-\Sigma s) + \exp(-\Sigma s) \int_0^s Q_l \exp(-\Sigma s') ds' \tag{72}$$

where $\varphi_{l0} = \varphi_l(0)$ denotes the incoming flux in the direction $\mathbf{\Omega}_l$ at the boundary of the computational cell. In the SC approach, the total cross-section Σ and the source Q_l are assumed to be constant in each computational cell. Therefore, the integration of equation (72) yields

$$\varphi_l(s) = \varphi_{l0} \exp(-\Sigma s) + \frac{Q_l}{\Sigma}(1 - \exp(-\Sigma s)) \tag{73}$$

Finite Element Methods

Finite element methods have been extensively used for the solution of neutron transport equation. They have been utilized in various different ways to solve different forms of the neutron transport equation. The finite element methods are generally based on weighted residual approaches which include Galerkin and Petrov-Galerkin methods, or variational approaches which include the classical Euler-Lagrange method, the method of bilinear functional and the generalized least square methods (see [2, 3, 4, 11, 18, 19, 23, 27, 29, 30, 34, 69, 58, 60, 52, 61, 75] and the references therein). Adaptivity [75], mixed and

41

mixed-hybrid formulations [29, 40] have been investigated. For the spatial discretization of the angularly discretised form of the transport equation (see (47) or (51)) by the finite element method, the spatial domain is partitioned into a finite set of non overlapping elements (which are generally intervals in 1-D, triangular and quadrilateral elements in 2-D, tetrahedral, triangular prisms and hexahedral elements in 3-D). Over each element, the flux is expanded in terms of interpolating basis functions associated to collocation points which generally positioned at the boundary of the element (corner, mid-edges or mid-faces). The finite element approximation can be continuous or discontinuous across the element boundaries. For the illustration of the finite element method, the weighted residual approach is applied to l^{th} equation of system (47) stated as follows:

$$\mathbf{\Omega}_l \cdot \nabla_x \varphi_l(\mathbf{x}) + \Sigma(\mathbf{x})\varphi_l(\mathbf{x}) = Q_l(\mathbf{x}), \tag{74}$$

The associated boundary condition is given by:

$$\varphi_l(\mathbf{x}) = u_l(\mathbf{x}), \quad \mathbf{n}(\mathbf{x}) \cdot \mathbf{\Omega}_l < 0, \quad x \in \partial D. \tag{75}$$

The solution of problem (74)-(75) belongs to the space [86]

$$H_l(D) = \{u \in L^2(D) \mid \mathbf{\Omega}_l \cdot \nabla_x u \in L^2(D)\}. \tag{76}$$

Continuous Finite Element Methods

Multiplying (74) by a trial function $\psi \in H_l(D)$ and integrating over the spatial domain, yields the following Galerkin weak form of the equation (74) :

Find $\varphi_l \in H_l(D)$, such that:

$$(\mathbf{\Omega}_l \cdot \nabla_x \varphi_l(\mathbf{x}) + \Sigma\varphi_l, \psi) = (Q_l, \psi), \quad \psi \in H_l(D), \tag{77}$$

where $(f, g) = \int_D fg d\mathbf{x}$. The boundary condition (75) can be incorporated into the weak formulation (78) using integration by parts. This method is known to produce spurious oscillation, particularly in the case of discontinuous solution [69]. The stability can be achieved by applying the Streamline Upwind Petrov-Galerkin (SUPG) methods (see [69, 27]). The SUPG formulation of (74) is the following :

Find $\varphi_l \in H_l(D)$, such that:

$$(\mathbf{\Omega}_l \cdot \nabla_x \varphi_l(\mathbf{x}) + \Sigma\varphi_l, \psi + \delta\mathbf{\Omega}_l \cdot \nabla_x \psi) = (Q_l, \psi + \delta\mathbf{\Omega}_l \cdot \nabla_x \psi), \quad \psi \in H_l(D), \tag{78}$$

The cell-wise constant parameter function δ depends on cell width and the total cross-section σ. Again, boundary condition can be recovered using integration by parts.

Discontinuous Finite Element Methods

We consider here the discontinuous Galerkin formulation of (74). Let $\mathbb{T}_h$ be a subdivision of D into a finite set of open and disjoint elements τ_j ($1 \leq j \leq N_h$) such that $\bigcup_{j=1}^{N_h} \tau_j = D$. The elements τ_j ($1 \leq j \leq N_h$) are assumed to be shape-regular. For an element τ, we define the finite element space

$$V_h(K) = \left\{ \psi \in L^2(K) \mid \psi_{|\tau} \in P_\tau^k, \ \tau \in \mathbb{T}_h \right\} \tag{79}$$

where P_τ^k is the set of polynomial of degree k defined on τ. The weak formulation of the discontinuous Galerkin method reads :

Find $V_h(K)$ such that :

$$(\varphi_l^h, -\boldsymbol{\Omega}_l \cdot \nabla_x \psi^h)_\tau + \langle \varphi_l^{h,n}, \psi^{h,-}\rangle_{\partial\tau_+} - \langle \varphi_l^{h,n}, \psi^{h,+}\rangle_{\partial\tau_-} + (\Sigma \varphi_l^h, \psi^h)_\tau = (Q_l, \psi^h)_\tau, \ \psi^h \in V_h(K), \tag{80}$$

where,

$$(f,g)_\tau = \int_{\partial\tau_\pm} fg\,d\mathbf{x}, \qquad \langle f,g\rangle_{\partial\tau_\pm} = \int_{\partial\tau_\pm} |\boldsymbol{\Omega}_l \cdot n(\mathbf{x})| fg\,ds$$

$$f^{h,+} = \lim_{s\to 0+} f(\mathbf{x}+s\boldsymbol{\Omega}_l) \qquad f^{h,-} = \lim_{s\to 0-} f(\mathbf{x}+s\boldsymbol{\Omega}_l)$$

$$\varphi_l^{h,n} = \varphi_l^{h,-} \ \text{ on } \ \partial\tau\backslash\partial D_- \qquad \varphi_l^{h,n} = u_l(\mathbf{x}) \ \text{ on } \ \partial D_-.$$

Nodal Methods

The nodal method approximates the multidimensional transport equation by a coupled system of 1-D transport equations. Efficient discretization methods for 1-D problems can then be utilized (see [49]). For each spatial coordinate, the 1-D equation is obtained by integrating the angularly discretised equation over the transverse coordinates. For instance for the discrete direction $\boldsymbol{\Omega}_l$, by integrating over y and z in the cell V_{ijk}, the following equation is obtained for $x_{i-1} \leq x \leq x_{i+1}$:

$$\mu_l \frac{d}{dx}\varphi_{ly_jz_k}(x) + \Sigma_{ijk}\varphi_{ly_jz_k}(x) = Q_{ly_jz_k}(x) - \left[\frac{\eta_l}{\Delta y_j}\varphi_{ly_{j+1/2}z_k}(x) + \frac{\xi_l}{\Delta z_k}\varphi_{ly_jz_{k+1/2}}(x)\right] \tag{81}$$

where

$$\varphi_{ly_jz_k}(x) = \tfrac{1}{\Delta y_j \Delta z_k} \int_{y_{j-1}}^{y_j} \int_{z_{k-1}}^{z_k} \varphi_l(x,y,z)\,dy\,dz,$$

$$\varphi_{ly_{j+1/2}z_k}(x) = \tfrac{1}{\Delta z_k} \int_{z_{k-1}}^{z_k} [\varphi_l(x,y_j,z) - \varphi_l(x,y_{j-1},z)]\,dz,$$

$$\varphi_{ly_jz_{k+1/2}}(x) = \tfrac{1}{\Delta y_j} \int_{y_{j-1}}^{y_j} [\varphi_l(x,y,z_k) - \varphi_l(x,y,z_{k-1})]\,dy,$$

$$Q_{ly_jz_k}(x) = \tfrac{1}{\Delta y_j \Delta z_k} \int_{y_{j-1}}^{y_j} \int_{z_{k-1}}^{z_k} Q_l(\mathbf{x})\,dy\,dz$$

43

Proceeding similarly for the spatial coordinates y and z, we obtain the following

$$\eta_l \frac{d}{dy} \varphi_{lx_i z_k}(y) + \Sigma_{ijk} \varphi_{lx_i z_k}(y) = Q_{lx_i z_k}(y) - \left[\frac{\mu_l}{\Delta x_i} \varphi_{lx_{i+1/2} z_k}(y) + \frac{\xi_l}{\Delta z_k} \varphi_{lx_i z_{k+1/2}}(y) \right]$$
(82)

$$\xi_l \frac{d}{dz} \varphi_{lx_i y_j}(z) + \Sigma_{ijk} \varphi_{lx_i y_j}(z) = Q_{lx_i y_j}(z) - \left[\frac{\mu_l}{\Delta x_i} \varphi_{lx_{i+1/2} y_j}(z) + \frac{\eta_l}{\Delta y_j} \varphi_{lx_i y_{j+1/2}}(z) \right]$$
(83)

The equations (81), (82) and (83) are exact in the sense that they contain no spatial approximations. These equations can be closed by taking:

$$\varphi_{ly_{j+1/2} z_k}(x) = \varphi_{lx_i z_k}(y_j) - \varphi_{lx_i z_k}(y_{j-1}), \quad \varphi_{ly_j z_{k+1/2}}(x) = \varphi_{lx_i y_j}(z_k) - \varphi_{lx_i y_j}(z_{k-1}),$$
$$\varphi_{lx_{i+1/2} z_k}(y) = \varphi_{ly_j z_k}(x_i) - \varphi_{ly_j z_k}(x_{i-1}), \quad \varphi_{lx_i z_{k+1/2}}(y) = \varphi_{lx_i y_j}(z_k) - \varphi_{lx_i y_j}(z_{k-1}),$$
$$\varphi_{lx_{i+1/2} y_j}(z) = \varphi_{ly_j z_k}(x_i) - \varphi_{ly_j z_k}(x_{i-1}), \quad \varphi_{lx_i y_{j+1/2}}(z) = \varphi_{lx_i z_k}(y_j) - \varphi_{lx_i z_k}(y_{j-1}).$$

2.4 Iterative Approaches

Due to the number of independent variables of the transport problem, the algebraic system equations resulting from the discretization of this problem is extremely large and thus difficult to invert directly. The solution strategy for solving this system of equations has then focussed on iterative methods [47]. Iterative methods are practical choices for the solution of the neutron transport equation. The general solution approach consists of three level of tasks:

1. A loop that computes the flux distribution in each energy group. When there is no up-scatter of neutrons, this loop consists of only one iteration. Otherwise, Jacobi and Gauss-Seidel like algorithm can be utilized.

2. A loop that computes the flux distribution for each time step. It consists of only one iteration for forward and backward time stepping.

3. An inner loop that solves the single-group steady state problem.

In this section, we give an outline of iterative methods for solving single-group steady state neutron transport equation. A comprehensive review on iterative methods for discrete ordinates particle transport calculations is given in [5].

The standard method is the source iteration method based on a decoupling between the differential and integral parts of the transport operator. This method becomes extremely slow in the critical case (optically thick and scattering dominant regions). Several acceleration techniques of the convergence of the source iteration method such as Diffusion Synthetic Acceleration (DSA) [5, 84], Transport Synthetic Acceleration (TSA) [5], Coarse Mesh Rebalance (CMR) [88], Quasi-Diffusion acceleration [87] and multigrid algorithms have been introduced and studied [30, 56, 5, 46, 63]. Alternative methods to the source iteration approach are the Krylov subspace iteration methods such as Conjugate Gradirnt (CG), Generalized Minimal Residual (GMRES), Bi-conjugate Gradient Stabilized (BiCGSTAB) and their preconditioned versions [5, 47, 66]. The Distribution Iteration (DI) method based on reducing the global transport equation into coupled-cell partial current that can be solved directly [36], have been applied. The angle space distribution iteration method because which combines a non-linear, high angular-resolution flux approximation within individual spatial cells with a coarse angular-resolution flux approximation that couples all cells in a spatial mesh, has proved its efficiency for slab geometry problems [83]. In [62], a new iterative method based on the idea of dividing the transport solution into its particular and homogeneous components was successfully implemented in slab geometry with isotropic scattering and one energy group. Based on the natural splitting of the integral part of transport operator, other methods such as Jacobi, Gauss-Seidel (see [79]) and Successive overrelaxation (SOR) iteration for the continuous problem, have been introduced and analyzed. They have been successfully applied to transport problem. Using the same splitting, an adaptation to the infinite dimensional case of the minimal residual iteration method (see [6, 7]) has been proposed for the solution of the transport in slab geometry, in 2-D cartesian geometry and in 1-D spherical geometry. This method has been proved to be efficient and it competes with the SOR method. Further, its preconditioned versions have been analyzed (see [80]).

2.5 Summary

In this chapter we gave an overview of deterministic methods for the solution of the neutron transport equation. Discretization methods for energy, time, angular and spatial variables have been briefly discussed. The discretization of these variables leads to a huge algebraic system of equations which are very difficult to solve directly. Iterative approaches for solving the transport equations have been briefly presented.

Chapter 3

New Splitting Iterative Methods for Solving Neutron Transport Equations

We present in this chapter a new class of iterative methods for solving the neutron transport equation. These methods are two step Alternating Direction Implicit (ADI) like iterative methods for the operator admitting Positive definite and m-Accretive Splitting (PAS). The iterations alternate between the positive definite and m-accretive parts of the operator. Theoretical analysis shows that the methods converge unconditionally to the solution of the equation (see [12]). Additionally, the analysis of a successive overrelaxation acceleration of this method yields convergence results similar to those obtained in presence of finite dimensional linear systems with matrices possessing property A (see [50, 89]). In the particular case which often arises in transport theory, where the positive definite part of the operator is self-adjoint, we obtain the Self-adjoint and m-Accretive Splitting (SAS) iterative method. Upper bound for the contraction factor of the SAS method is derived and the convergence of its incomplete version is investigated. Further, the SAS iteration leads to a fixed point problem where the operator is a 2 by 2 matrix of operators. An infinite dimensional adaptation of a minimal residual algorithm is then applied to solve the resulting matrix operator equation. The theoretical convergence results of the method are given. The convergence of the symmetric Gauss-Seidel and polynomial preconditioning of the minimal residual algorithm is established. The upper bounds for the rate of residual decreasing of these minimal residual methods are derived.

The convergence of the proposed methods is illustrated by numerical examples in which integro-differential problems of transport theory are considered.

3.1   Introduction

Iterative methods are intensively used for the solution of the transport equation (see [5, 6, 7, 34, 30, 56, 58, 66, 67, 77, 79, 85, 84] and the reference therein). The standard method is the source iteration method based on a decoupling between the differential and integral parts of the transport operator. Based on the natural splitting of the integral part of transport operator, other methods such as Jacobi, Gauss-Seidel (see [79]) and Successive overrelaxation (SOR) iteration have been successfully applied to transport problem. Using the same splitting, an adaptation to the infinite dimensional case of the minimal residual iteration method (see [6, 7]) has been proposed for the solution of the transport in slab geometry, in 2-D cartesian geometry and in 1-D spherical geometry. This method has been proved to be efficient and it competes with the SOR method. Further, its preconditioned versions have been analyzed (see [80]).

In this chapter, focus is given on iterative methods for the numerical treatment of the single group steady state neutron transport equation in slab geometry, bounded convex domain of $\mathbb{R}^n$ $(n = 2, 3)$ and in 1-D spherical geometry.

We introduce a generic ADI-like iterative method (see [59]) based on positive definite and m-accretive splitting (PAS) for linear operator equations with operators admitting such splitting. As mentioned above, theoretical results show the convergence of the method and its SOR acceleration (see [12]) yields convergence results similar to those obtained in presence of finite dimensional systems with matrices possessing property A (see [50, 89]). The proposed methods are illustrated by a numerical example in which an integro-differential problem of transport theory is considered. In the particular case where the positive definite part of the linear equation operator is self-adjoint, an upper bound for the contraction factor of the iterative method which depends solely on the spectrum of the self-adjoint part is derived (see [13]) . As such, this method has been successfully applied to the neutron transport equation in slab and 2-D cartesian geometry (see [13]) and in 1-D spherical geometry (see [15]).

The self-adjoint and m-accretive splitting leads to a fixed point problem where the operator is a 2 by 2 matrix of operators. An infinite dimensional adaptation of minimal residual and preconditioned minimal residual algorithms using Gauss-Seidel, symmetric Gauss-Seidel and polynomial preconditioning is then applied to solve the matrix operator equation. Theoretically, the methods are shown to be unconditionally convergent and upper bounds of the rate of residual decreasing which depend solely on the spectrum of the self-adjoint part of the operator are derived (see [16, 17]).

The convergence of theses solvers is numerically illustrated on a sample neutron transport problem in 2-D geometry. Various test cases, including pure scattering and optically thick domains are considered.

The remainder of this chapter is structured as follows:

In section 3.3 we present preliminaries results and give the mathematical setting of the problem.

In section 3.3 we present the PAS iterative method. The convergence analysis of the method and its SOR acceleration is provided. The convergence of the proposed methods are illustrated by a numerical example in which an integro-differential problem of transport theory is considered.

In section 3.4 we present the convergence analysis of the SAS iterative method and its incomplete version. The convergence of the method is numerically illustrated and compared with the standard Source Iteration method and multigrid method on sample problems in slab geometry and in two dimensional space.

In section 3.5 we introduce and analyze an infinite dimensional adaptation of a minimal residual algorithm linked to the self-adjoint and m-Accretive Splitting. Comparative numerical results are presented for a sample neutron transport problem in 2-D geometry.

In section 3.6 the convergence of the previous minimal residual algorithm with Symmetric Gauss-Seidel and polynomial preconditioning is established and comparative numerical results are presented.

The section 3.7 is devoted to some concluding remarks.

3.2   Preliminaries and Mathematical Setting

3.2.1   Preliminaries

Let us consider a Hilbert space H with inner product $(.,.)_H$ and the associated norm $\|.\|_H$. Let A be an unbounded linear operator on H with domain $\mathcal{D}(A)$. Let I denote the identity operator on H.

Definition 3.2.1. An operator $A : D(A) \subset H \rightarrow H$ is said to be m-accretive if $\forall u \in D(A)$, $(Au, u)_H \geq 0$ and $\forall q \in H$, $\exists u \in D(A)$ such that $Au + u = q$.

We have the following results (see [25, 34]) :

Theorem 3.2.1. *Assume that A is an m-accretive operator on H. Then:*

1. *$D(A)$ is dense in H.*

2. *The operator A is closed.*

3. *$\forall \alpha > 0$, $(I + \alpha A)$ is bijective from $D(A)$ to H, the operator $(I + \alpha A)^{-1}$ is bounded and $\|(I + \alpha A)^{-1}\|_H \leq 1$.*

Corollary 3.2.2. *If A is an m-accretive operator on H, then for any positive constant α, the operator $(\alpha I + A)^{-1}$ is bounded on H.*

Proof. If the operator A is m-accretive on H, then for any positive constant α we have

$$\|(I + \alpha A)^{-1}\|_H = \frac{1}{\alpha}\|(\alpha I + A)^{-1}\|_H.$$

Thus $(\alpha I + A)^{-1}$ is bounded on H (Theorem 3.2.1). $\qquad \square$

Theorem 3.2.3. *If A is an m-accretive operator in the Hilbert space H, then for $\alpha > 0$,*

$$\|(\alpha I - A)(\alpha I + A)^{-1}\| \leq 1. \tag{1}$$

Proof. Let $\alpha > 0$. If the operator A is m-accretive then

$$\|(\alpha I - A)u\|^2 - \|(\alpha I + A)u\|^2 = -4\alpha(Au, u) \leq 0.$$

Taking $u = (\alpha I + A)^{-1}\varphi, (\varphi \in H)$, we obtain

$$\left\|(\alpha I - A)(\alpha I + A)^{-1}\varphi\right\|^2 \leq \|\varphi\|^2.$$

It follows that $\|(\alpha I - A)(\alpha I + A)^{-1}\| \leq 1$. $\qquad \square$

Let A be an m-accretive operator on H. We consider in $\mathcal{D}(A)$ the norm

$$\|u\|_{\mathcal{D}(A)} = \left(\|u\|_H^2 + \|Au\|_H^2\right)^{\frac{1}{2}}. \tag{2}$$

Theorem 3.2.4. *Let α be a positive constant. The functional $\rho_{A(\alpha)}$ defined on $\mathcal{D}(A)$ by*

$$\rho_{A(\alpha)}(u) = \|(\alpha I + A)u\|_H, \tag{3}$$

is a norm on $\mathcal{D}(A)$ equivalent to $\|.\|_{\mathcal{D}(A)}$.

Proof. Let α be a positive constant. It can be easily seen from the linearity of the operator $\alpha I + A$ and the properties of the norm $\|.\|_H$ that $\forall u, v \in \mathcal{D}(A)$ and $\beta \in \mathbb{C}$

$$\rho_{A(\alpha)}(u + v) \leq \rho_{A(\alpha)}(u) + \rho_{A(\alpha)}(v) \text{ and } \rho_{A(\alpha)}(\beta u) = |\beta|\rho_{A(\alpha)}(u).$$

Moreover, since A is m-accretive, $\alpha I + A$ is positive definite and $\rho_{A(\alpha)}(u) = 0$ if and and only if $u = 0$. It then follows that $\rho_{A(\alpha)}$ is a norm on $\mathcal{D}(A)$.

Let $u \in \mathcal{D}(A)$. We have

$$\left(\rho_{A(\alpha)}(u)\right)^2 = \alpha^2\|u\|_H^2 + \|Au\|_H^2 + 2\alpha(u, Au)_H.$$

Since $(Au, u)_H \geq 0$ (A is accretive) and $(Au, u)_H \leq \frac{\|u\|_H^2 + \|Au\|_H^2}{2}$, we have

$$\min\{\alpha^2, 1\}\|u\|_{\mathcal{D}(A)}^2 \leq \left(\rho_{A(\alpha)}(u)\right)^2 \leq (\alpha + 1)^2\|u\|_{\mathcal{D}(A)}^2.$$

It follows that, the norms $\|.\|_{\mathcal{D}(A)}$ and $\rho_{A(\alpha)}$ are equivalent in $\mathcal{D}(A)$. $\qquad\square$

In the following, the norm $\rho_{A(\alpha)}$ is denoted by $\|.\|_{A(\alpha)}$.

3.2.2 Mathematical Setting

The neutron transport equation in cartesian geometry

The general single group, steady state first order neutron transport equation reads:

$$\mathbf{\Omega} \cdot \nabla_x \psi(x, \mathbf{\Omega}) + \sigma(x)\psi(x, \mathbf{\Omega}) = \int_{S_2} \kappa(x, \mathbf{\Omega}, \mathbf{\Omega}')\psi(x, \mathbf{\Omega})d\mathbf{\Omega}' + q(x, \mathbf{\Omega}), \tag{4}$$

where $\sigma(x)$ is the total cross section; $\kappa(x, \mathbf{\Omega}, \mathbf{\Omega}')$ is a positive kernel specifying the scattering of particles; $q(x, \mathbf{\Omega})$ is a known particles source and $\psi(x, \mathbf{\Omega})$ represents the angular flux to be determined for all point $x = (x_1, x_2, x_3)$ in a bounded convex region $D \subset R^3$ with a sufficiently smooth boundary ∂D and all possible travel directions $\mathbf{\Omega} = (\theta, \varphi)$ on the unit sphere S_2. The boundary condition prescribing the inflow of particles into the region D reads:

$$\psi(x, \mathbf{\Omega}) = u_0(x, \mathbf{\Omega}), \quad \forall(x, \mathbf{\Omega}) \in \partial D_-, \tag{5}$$

51

where $\partial D_- = \{(x, \mathbf{\Omega}) \in \partial D \times S_2, \ \mathbf{\Omega}.\vec{n}(x) < 0\}$, $\vec{n}(x)$ being the outer unit normal to D at point $x \in \partial D$. When $\frac{\partial \psi(x)}{\partial x_3} = 0$, the problem (4)-(5) reduces to a 2D-problem in spatial space. In that case, the third component of the inner product $\mathbf{\Omega} \cdot \nabla_x \psi$ is ignored.

In the case of slab geometry it is assumed that $\frac{\partial \psi}{\partial x_1} = \frac{\partial \psi}{\partial x_2} = 0$. Defining $\mu = cos(\theta)$, where θ denotes the angle between $\mathbf{\Omega}$ and the $z-$axis, the angular flux becomes $\psi(x, \mathbf{\Omega}) \equiv \psi(z, \mu)$ and problem (4)-(5) reduces to

$$\begin{cases} \mu \frac{\partial \psi}{\partial z} + \sigma(z)\psi - \int_{-1}^{1} \kappa(z, \mu, \mu')\psi(z, \mu')d\mu' = q(z, \mu) \\ \qquad\qquad \psi(z_l, \mu) = g_l(\mu) \quad \text{for } \mu > 0 \\ \qquad\qquad \psi(z_r, \mu) = g_r(\mu) \quad \text{for } \mu < 0 \end{cases} \qquad (6)$$

(see [34, 58, 61]). In this case, we set $D =]z_l, z_r[$ and $S_2 = [-1, 1]$. Without loss of generality, we assume in the following vacuum boundary conditions, which means that $u_0(x, \mathbf{\Omega}) \equiv 0$ in (5) and $g_l(\mu) \equiv g_r(\mu) \equiv 0$ in (62).

Let $Q = D \times S_2$. We define the spaces

$$W^2(Q) = \left\{ \psi \in L^2(Q), \ \mathbf{\Omega} \cdot \nabla_x \psi \in L^2(Q) \right\}, \qquad (7)$$

and

$$W_0 = \left\{ \psi \in W^2(Q), \psi(x, \mathbf{\Omega}) = 0, \ (x, \mathbf{\Omega}) \in \partial D_- \right\}. \qquad (8)$$

The space $W^2(Q)$ is equipped with the norm

$$\|\psi\|_{W^2} = \left(\|\psi\|^2 + \|\mathbf{\Omega} \cdot \nabla_x \psi\|^2 \right)^{\frac{1}{2}}, \qquad (9)$$

where $\|.\|$ denotes the usual $L^2(Q)$ norm. The functions of $W^2(Q)$ have traces on ∂D_- in $L^2(\partial D \times S_2)$ (see [34]).

Let

$$\begin{cases} K\psi(x, \mathbf{\Omega}) = \int_{S_2} \kappa(x, \mathbf{\Omega}, \mathbf{\Omega}')\psi(x, \mathbf{\Omega}')d\mathbf{\Omega}' \\ A\psi(x, \mathbf{\Omega}) = \mathbf{\Omega} \cdot \nabla_x \psi(x, \mathbf{\Omega}) \\ \Sigma\psi(x, \mathbf{\Omega}) = \sigma(x)\psi(x, \mathbf{\Omega}) \end{cases} \qquad (10)$$

We have

Theorem 3.2.5. *The operator A is m-accretive in $L^2(Q)$.*

Proof. See [34]. $\qquad\qquad\qquad\qquad\qquad\qquad\qquad\qquad\qquad\qquad\qquad\qquad\quad\square$

Assuming that:

$(A1)$ $\sigma \in L^\infty(D)$, $\exists \sigma_0 > 0$ such that $\sigma(x) \geq \sigma_0$ a.e. on D;

$(A2)$ $\kappa(x, \boldsymbol{\Omega}, \boldsymbol{\Omega'}) \geq 0$;

$(A3)$ $\kappa(x, \boldsymbol{\Omega}, \boldsymbol{\Omega'}) = \kappa(x, \boldsymbol{\Omega'}, \boldsymbol{\Omega})$;

$(A4)$ $\exists c \in [0, 1)$, $\displaystyle\int_{S_2} \kappa(x, \boldsymbol{\Omega}, \boldsymbol{\Omega'}) d\boldsymbol{\Omega'} \leq \sigma_0 c$ a.e. on Q,

the following results hold (see [6, 34, 61]) :

Theorem 3.2.6. *The operators Σ and K satisfy the following properties:*

1. *The operator Σ is self-adjoint and positive definite.*

2. *The operator K is self-adjoint. If in addition the function κ is positive, then K is positive definite*

3. *The operator $\Sigma - K$ is self-adjoint and positive definite.*

Setting

$$T = A + \Sigma - K, \tag{11}$$

the equation (4) may be written as

$$T\psi(x, \boldsymbol{\Omega}) = q(x, \boldsymbol{\Omega}). \tag{12}$$

We have

$$\mathcal{D}(K) = \mathcal{D}(\Sigma) = L^2(Q), \tag{13}$$

where $\mathcal{D}(B)$ denotes the domain of operator B. It follows that

$$\mathcal{D}(T) = \mathcal{D}(A) = W_0. \tag{14}$$

Since A is m-accretive and $\Sigma - K$ is positive definite, we have for $\psi \in W_0$, we have

$$(A\psi, \psi) \geq 0 \text{ and } ((\Sigma - K)\psi, \psi) > a\|\psi\|^2$$

where a is a positive constant. Therefore $(T\psi, \psi) > a\|\psi\|^2$ and T is positive definite. Thus (see [34])

Theorem 3.2.7. *Under the assumptions $(A1)$–$(A4)$, the solution of problem (4)-(5) exists and is unique in W_0.*

53

The neutron transport equation in 1-D spherical geometry

The single group steady state first order neutron transport equation in 1-D spherical geometry reads:

$$\mu\frac{\partial u}{\partial r} + \frac{1-\mu^2}{r}\frac{\partial u}{\partial \mu} + \sigma u = \int_{-1}^{1}\kappa(r,\mu,\mu')u(r,\mu')d\mu' + q(r,\mu) \quad \text{in } (0,R)\times(-1,1), \quad (15)$$

where $u(r,\mu)$ is the neutron flux. The region occupied by the particles is a sphere of radius $R > 0$; r is the distance from the center of the sphere; μ is the cosine of the angle the particle velocity makes with the radius; $\sigma(r)$ is the total cross section; $\kappa(r,\mu,\mu')$ is a positive kernel specifying the scattering of particles; $q(r,\mu)$ is a known particles source and $u(r,\mu)$ represents the angular flux to be determined for all point $r \in (0,R)$ and all $\mu \in (-1,1)$. The boundary conditions prescribing the inflow of particles into the sphere reads:

$$u(R,\mu) = 0 \text{ for } \mu \in (-1,0). \quad (16)$$

Let $\Omega = (0,R)\times(-1,1)$. We define the space

$$W^2(\Omega) = \left\{ u \in L^2(\Omega), \mu\frac{\partial u}{\partial r} \in L^2(\Omega) \text{ and } \frac{1-\mu^2}{r}\frac{\partial u}{\partial \mu} \in L^2(\Omega)\right\}, \quad (17)$$

endowed with the norm $\|u\|_{W^2}^2 = \|u\|^2 + \left\|\mu\frac{\partial u}{\partial r} + \frac{1-\mu^2}{r}\frac{\partial u}{\partial \mu}\right\|^2$, where $\|.\|$ denotes the standard $L^2(\Omega)$ norm. The functions in $W^2(\Omega)$ have traces on $\{0\}\times(-1,1)$ and $\{R\}\times(-1,1)$ which lie in in $L^2(\{0\}\times(-1,1))$ and $L^2(\{R\}\times(-1,1))$, respectively (see [34]).

Let A, K, Σ, and T be the operators defined in $W^2(\Omega)$ by:

$$Au = \mu\frac{\partial u}{\partial r} + \frac{1-\mu^2}{r}\frac{\partial u}{\partial \mu}, \quad Ku = \int_{-1}^{1}\kappa(r,\mu,\mu')u(r,\mu')d\mu', \quad \Sigma u = \sigma(r)u \quad (18)$$

and

$$Tu(r,\mu) = Au(r,\mu) + \Sigma u(r,\mu) - Ku(r,\mu). \quad (19)$$

Having this in mind, the problem (15)-(16) is stated as follows:

$$Tu(r,\mu) = q(r,\mu). \quad (20)$$

We have

$$\mathcal{D}(K) = \mathcal{D}(\Sigma) = L^2(\Omega). \quad (21)$$

It follows that

$$\mathcal{D}(T) = \mathcal{D}(A) = \left\{u \in W^2(\Omega); u(R,\mu) = 0 \text{ for } \mu < 0\right\}. \quad (22)$$

54

Proceeding as in [6, 7, 13], we make the following assumptions:

$(a1)$ $\sigma \in L^{\infty}((0, R))$, $\exists \sigma_0 > 0$ such that $\sigma(x) \geq \sigma_0$ a.e. on $(0, R)$.

$(a2)$ $\kappa(r, \mu, \mu') = \kappa(r, \mu', \mu)$ and κ is positive .

$(a3)$ $\exists c \in [0, 1), \displaystyle\int_{-1}^{1} \kappa(r, \mu, \mu')d\mu' \leq \sigma_0 c$ a.e. on Ω.

Under the above assumptions, the results of Theorem 3.2.6 hold (see [6, 34, 61]) .

3.3   Positive Definite and m-Accretive Splitting Iterative Method

Linear equations with operators admitting positive definite and m-accretive splitting often arise in physics and in engineering. They find many applications in particle transport, radiative transfer, diffusion convection etc... In this section, we present a two step iterative method for the solution of a linear equation, where the operator admits such splitting. The iterations alternate between the positive definite and m-accretive part of the operator.

Let us consider a Hilbert space H with inner product $(.,.)$ and norm $\|.\|$ and let T be a linear operator on H with domain $\mathcal{D}(T)$ and range $\mathcal{R}(T) = H$. We denote by I the identity operator. Suppose that we need to solve in $\mathcal{D}(T)$, the following problem

$$Tu = q, \tag{23}$$

where $q \in H$ is given and $u \in \mathcal{D}(T)$ is the unknown.

We assume that the operator T admits the following splitting:

$$T = P + A, \tag{24}$$

where P is a bounded positive definite operator and A is an m-Accretive operator. Therefore, the operator T is positive definite and equation (23) admits a unique solution in H.

We introduce in this section a two-step iteration method linked to the Positive definite and m-Accretive Splitting (24) of operator T. We investigate the convergence of this iterative method. Theoretical analysis shows that this iterative method converges to the solution of equation (23). An analysis of a successive overrelaxation acceleration of this iteration method is provided. We obtain similar results as in the case of finite dimensional linear systems with coefficient matrices possessing Property A (see [21, 50]). The proposed methods are illustrated by a numerical example in which an integro-differential problem of transport theory is considered.

Assuming (23), we consider in $\mathcal{D}(T)$ the norm

$$\|u\|_{\mathcal{D}(T)} = \left(\|u\|^2 + \|Au\|^2\right)^{\frac{1}{2}}. \tag{25}$$

Since the operator A is m-accretive and $\mathcal{D}(T) = \mathcal{D}(A)$, we deduce from Theorem 3.2.4 that for any positive constant α, the norms $\|.\|_{\mathcal{D}(T)}$ and $\|.\|_{A(\alpha)}$ are equivalent.

The outline of this section is as follows: in subsection 3.3.1, we present the two step iterative methods and the convergence analysis. Subsection 3.3.2 deals with the analysis of a successive overrelaxation acceleration of the method. The numerical results are presented in subsection 3.3.3.

3.3.1   The iteration method

We present in this section an iterative method for the solution of the problem (23). It is a two-step iteration method relies on the splitting (24) and alternating between the positive definite and the m-accretive parts of the operator T.

Let α be a positive constant. The following two-step splitting is obtained from (24):

$$\begin{cases} T = (\alpha I + P) - (\alpha I - A) \\ T = (\alpha I + A) - (\alpha I - P) \end{cases}. \tag{26}$$

The two-step splitting (26) leads to the following Positive definite and m-Accretive Splitting (PAS) iteration method:

Given an initial guess $u^{(0)} \in D(T)$, for $k = 0, 1, \ldots$ until $\{u^{(k)}\}$ converges, calculate

$$\begin{cases} (\alpha I + P)u^{(k+\frac{1}{2})} = (\alpha I - A)u^{(k)} + q \\ (\alpha I + A)u^{(k+1)} = (\alpha I - P)u^{(k+\frac{1}{2})} + q \end{cases}. \tag{27}$$

From equation (27), we deduce that u^{k+1} satisfies

$$(\alpha I + A)u^{(k+1)} = M(\alpha)(\alpha I + A)u^{(k)} + N(\alpha)q, \tag{28}$$

where

$$M(\alpha) = V(\alpha)U(\alpha) \text{ and } N(\alpha) = 2\alpha(\alpha I + P)^{-1}; \tag{29}$$

56

with

$$V(\alpha) = (\alpha I - P)(\alpha I + P)^{-1} \text{ and } U(\alpha) = (\alpha I - A)(\alpha I + A)^{-1}. \tag{30}$$

Therefore, the exact solution u^* of the problem (23) verifies

$$\|u^{(k+1)} - u^*\|_{A(\alpha)} \le \|M(\alpha)\| \|u^{(k)} - u^*\|_{A(\alpha)}. \tag{31}$$

It is well known that the iteration method (27) converge (in the sense of the norm $\|.\|_{A(\alpha)}$) if the operator $M(\alpha)$ is a strict contraction or equivalently, if

$$\|(M(\alpha)\| < 1. \tag{32}$$

Theorem 3.3.1. *Convergence of the PAS iteration method.*

Let α be a positive constant. The norm $\|(M(\alpha)\|$ of the iteration operator $M(\alpha)$ verifies

$$\|(M(\alpha)\| < 1. \tag{33}$$

Therefore it holds that, the PAS iteration converges to the unique solution $u^* \in \mathcal{D}(T)$ of the problem (23).

Proof. The proof of this Theorem is based on the result of Theorem 3.2.3 and the following lemma.

Lemma 3.3.1. *If X is a positive definite operator in the Hilbert space H, then for $\alpha > 0$,*

$$\|(\alpha I - X)(\alpha I + X)^{-1}\| < 1. \tag{34}$$

Proof. The relation (34) is obtained by proceeding similarly as in the proof of Theorem 3.2.3. $\qquad\square$

We have

$$\|M(\alpha)\| \le \|U(\alpha)\| \, \|V(\alpha)\|.$$

Since P is positive definite and A is m-accretive, we deduce from Theorem 3.2.3 and Lemma 3.3.1 that

$$\|U(\alpha)\| = \|(\alpha I - A)(\alpha I + A)^{-1}\| \le 1$$

and

$$\|V(\alpha)\| = \|(\alpha I - P)(\alpha I + P)^{-1}\| < 1.$$

It then follows that

$$\|M(\alpha)\| \le \|U(\alpha)\| \, \|V(\alpha)\| < 1. \; \square$$

$\square$

From the equivalence between the norms $\|.\|_{\mathcal{D}(T)}$ and $\|.\|_{A(\alpha)}$ follows the convergence with respect to the norm $\|.\|_{\mathcal{D}(T)}$ in $\mathcal{D}(T)$. Additionally, since

$$\|u^{(k+1)} - u^*\| \le \|u^{(k+1)} - u^*\|_{\mathcal{D}(T)}; \; k = 0, 1, 2, \cdots , \tag{35}$$

we have

$$\lim_{k \to +\infty} \|u^{(k)} - u^*\| = \lim_{k \to +\infty} \|u^{(k)} - u^*\|_{\mathcal{D}(T)} = 0. \tag{36}$$

Thus the sequence $u^{(k)}$ converges in $\mathcal{D}(T)$ with respect to the norm $\|.\|$.

Each step of the PAS iterative method is constituted of two-half steps which require finding solutions of linear equations with operators $(\alpha I + P)$ and $(\alpha I + A)$. These linear operator equations can be solved approximately using appropriate methods with respect to the properties of each operators. This results in the inexact Positive definite and m-Accretive splitting (IPAS) iteration for solving the linear operator equation (23).

3.3.2   Successive overrelaxation (SOR) acceleration of the PAS Iteration Method.

The following fixed point equation can be derived from the definition of the PAS iteration (27):

$$\begin{cases} (\alpha I + P)u_1 = (\alpha I - A)u_2 + q \\ (\alpha I + A)u_2 = (\alpha I - P)u_1 + q \end{cases}. \tag{37}$$

In the operator form, the system (37) reads

$$\mathbf{T}(\alpha)\mathbf{u} = \mathbf{q}, \tag{38}$$

where the matrix of operators $T(\alpha)$ and the vector functions $\mathbf{u}$ and $\mathbf{q}$ are defined as follows

$$\mathbf{T}(\alpha) = \begin{pmatrix} (\alpha I + P) & -(\alpha I - A) \\ -(\alpha I - P) & (\alpha I + A) \end{pmatrix}, \quad \mathbf{u} = \begin{pmatrix} u_1 \\ u_2 \end{pmatrix} \text{ and } \mathbf{q} = \begin{pmatrix} q \\ q \end{pmatrix}.$$

Theorem 3.3.2. *Let α be a positive constant. The solution of linear operator equation (38) exits in $\mathcal{D}(T) \times \mathcal{D}(T)$ and is unique.*

58

Proof. For $\alpha > 0$, the operators $(\alpha I + A)$ and $(\alpha I + P)$ are positive definite. It follows that $(\alpha I + A)^{-1}$ and $(\alpha I + P)^{-1}$ exist and it holds that

$$\mathbf{T}(\alpha) = \begin{pmatrix} I & 0 \\ -(\alpha I - P)(\alpha I + P)^{-1} & I \end{pmatrix} \begin{pmatrix} (\alpha I + P) & -(\alpha I - A) \\ 0 & A(\alpha) \end{pmatrix}$$

with

$$\begin{aligned} A(\alpha) &= (\alpha I + A) - (\alpha I - P)(\alpha I + P)^{-1}(\alpha I - A) \\ &= [I - (\alpha I - P)(\alpha I + P)^{-1}(\alpha I - A)(\alpha I + A)^{-1}](\alpha I + A) \\ &= (I - M(\alpha))(\alpha I + A). \end{aligned}$$

Since $\|M(\alpha)\| < 1$, $(I - M(\alpha))^{-1}$ exists and $A(\alpha)$ is invertible. It then follows that the matrix operator $\mathbf{T}(\alpha)$ is invertible. $\square$

We consider in $H \times H$ the norm $\|\|.\|\|$ given for $\mathbf{u} = \begin{pmatrix} u_1 \\ u_2 \end{pmatrix} \in H \times H$ by

$$\|\|\mathbf{u}\|\|^2 = \|u_1\|^2 + \|u_2\|^2. \tag{39}$$

Theorem 3.3.3. *Let α be a positive constant. If u^* is the exact solution of problem (23), then $\mathbf{u}^* = \begin{pmatrix} u^* \\ u^* \end{pmatrix}$ is the exact solution of equation (38). Conversely, if $\mathbf{u}^* = \begin{pmatrix} u^* \\ v^* \end{pmatrix}$ is the exact solution of equation (38) then $u^* = v^*$ and, u^* is the exact solution of problem (23).*

Proof. Let u^* be the solution of problem (23). We have $Tu^* = q$ and

$$\mathbf{T}(\alpha) \begin{pmatrix} u^* \\ u^* \end{pmatrix} = \begin{pmatrix} (\alpha I + P)u^* - (\alpha I - A)u^* \\ -(\alpha I - P)u^* + (\alpha I + A)u^* \end{pmatrix} = \begin{pmatrix} Tu^* \\ Tu^* \end{pmatrix} = \mathbf{q}.$$

Let $\mathbf{u}^* = \begin{pmatrix} u^* \\ v^* \end{pmatrix}$ be the solution of equation (38). The functions u^* and v^* satisfy

$$(\alpha I + P)u^* - (\alpha I - A)v^* = q, \tag{40}$$

$$-(\alpha I - P)u^* + (\alpha I + A)v^* = q. \tag{41}$$

By subtracting (41) from (40), we have

$$2\alpha u^* - 2\alpha v^* = 0.$$

It follows that $v^* = u^*$. Substituting v^* in (40), we obtain $Tu^* = q$. $\square$

59

Let $\mathbf{P}(\alpha)$ be the matrix operator defined in $\mathcal{D}(T) \times \mathcal{D}(T)$ by:

$$\mathbf{P}(\alpha) = \begin{pmatrix} (\alpha I + P) & 0 \\ 0 & (\alpha I + A) \end{pmatrix}. \tag{42}$$

The preconditioning of the system (38) from the right by $[\mathbf{P}(\alpha)]^{-1}$ leads to the following system

$$\mathbf{T}_1(\alpha)\mathbf{u} = \mathbf{q} \tag{43}$$

where the matrix of operator $\mathbf{T}_1(\alpha)$ reads

$$\mathbf{T}_1(\alpha) = \begin{pmatrix} I & -A_1(\alpha) \\ -P_1(\alpha) & I \end{pmatrix}, \tag{44}$$

with $A_1(\alpha) = (\alpha I - A)(\alpha I + A)^{-1}$ and $P_1(\alpha) = (\alpha I - P)(\alpha I + P)^{-1}$. The solution $\mathbf{v}^*$ of problem (38) reads

$$\mathbf{v}^* = [\mathbf{P}(\alpha)]^{-1}\mathbf{u}^*, \tag{45}$$

where $\mathbf{u}^*$ is solution of (43).

Since all the operators of the matrix $\mathbf{T}_1(\alpha)$ are bounded in H, $\mathbf{T}_1(\alpha)$ is bounded in $H \times H$.

Given an initial guest $\mathbf{u}^{(0)}$ in $H \times H$, the Jacobi iteration for the solution of (43) reads

$$\mathbf{u}^{(k+1)} = \mathbf{J}(\alpha)\mathbf{u}^{(k)} + \mathbf{q}, \tag{46}$$

where

$$\mathbf{J}(\alpha) = \begin{pmatrix} 0 & A_1(\alpha) \\ P_1(\alpha) & 0 \end{pmatrix}. \tag{47}$$

The corresponding SOR iteration with the relaxation parameter ω, is

$$\mathbf{u}^{(k+1)} = \mathbf{L}_\omega(\alpha)\mathbf{u}^{(k)} + \mathbf{q}_\omega(\alpha), \tag{48}$$

where

$$\mathbf{L}_\omega(\alpha) = \begin{pmatrix} (1-\omega)I & \omega A_1(\alpha) \\ \omega(1-\omega)P_1(\alpha) & (1-\omega)I + \omega^2 M(\alpha) \end{pmatrix}. \tag{49}$$

The choice of $\omega = 1$ in the method (48) results in the Gauss-Seidel iteration for solving the system (43).

Theorem 3.3.4. *The Jacobi and Gauss-Seidel iteration methods for the solution of (43) converge.*

60

Proof. For $\mathbf{u} = \begin{pmatrix} u_1 \\ u_2 \end{pmatrix} \in H \times H,$

$$\||J(\alpha)\mathbf{u}\||^2 = \|A_1(\alpha)u_2\|^2 + \|P_1(\alpha)u_1\|^2 < \|u_2\|^2 + \|u_1\|^2 = \||\mathbf{u}\||^2.$$

It follows that $\||\mathbf{J}(\alpha)\|| < 1$ and the Jacobi iteration method (46) converges.

The Gauss-Seidel iteration is equivalent to the following iteration:

$$\begin{cases} u_1^{(k+1)} = A_1(\alpha)P_1(\alpha)u_1^{(k)} + (A_1(\alpha) + I)q \\ u_2^{(k+1)} = P_1(\alpha)A_1(\alpha)u_2^{(k)} + (P_1(\alpha) + I)q \end{cases}.$$

Let $\mathbf{u}* = \begin{pmatrix} u_1* \\ u_2* \end{pmatrix}$ be the exact solution to (43), we have

$$\begin{cases} u_1^{(k+1)} - u_1^* = A_1(\alpha)P_1(\alpha)[u_1^{(k)} - u_1^*] \\ u_2^{(k+1)} - u_2^* = P_1(\alpha)A_1(\alpha)[u_2^{(k)} - u_2^*] \end{cases}$$

and

$$\|u_1^{(k+1)} - u_1^*\| \leq \|A_1(\alpha)P_1(\alpha)\|\|u_1^{(k)} - u_1^*\| < \|u_1^{(k)} - u_1^*\|,$$
$$\|u_2^{(k+1)} - u_2^*\| \leq \|P_1(\alpha)A_1(\alpha)\|\|u_2^{(k)} - u_2^*\| < \|u_2^{(k)} - u_2^*\|.$$

Therefore, the exact solution $\mathbf{u}^*$ verifies

$$\||\mathbf{u}^{(k+1)} - \mathbf{u}^*\|| < \||\mathbf{u}^{(k)} - \mathbf{u}^*\||.$$

It follows that the Gauss-Seidel iteration method converges.

$\square$

Theorem 3.3.5. *Let R_1 and R_2 be two operators defined from H to H. If R_1 and R_2 are bounded, then for $\gamma \neq 0$, the operators*

$$\mathbf{R} = \begin{pmatrix} 0 & R_1 \\ R_2 & 0 \end{pmatrix} \quad and \quad \mathbf{R}(\gamma) = \begin{pmatrix} 0 & \gamma R_1 \\ \frac{1}{\gamma}R_2 & 0 \end{pmatrix}$$

have the same spectrum.

Proof. We have to prove that if $\lambda \notin \sigma(\mathbf{R})$, then $\lambda \notin \sigma(\mathbf{R}(\gamma))$ and conversely.

Let $\gamma \neq 0$ and $\lambda \in \mathbb{C}$. The operators R and $R(\gamma)$ satisfy

$$\mathbf{R} - \lambda\mathbf{I} = \mathbf{X}(\gamma)(\mathbf{R}(\gamma) - \lambda\mathbf{I})\mathbf{X}^{-1}(\gamma);$$
$$R(\gamma) - \lambda\mathbf{I} = \mathbf{X}^{-1}(\gamma)(\mathbf{R} - \lambda\mathbf{I})\mathbf{X}(\gamma).$$

where $\mathbf{X}(\gamma) = \begin{pmatrix} I & 0 \\ 0 & \frac{1}{\gamma}I \end{pmatrix}$. The operator $\mathbf{X}(\gamma)$ is bounded and has a bounded inverse. If $\lambda \notin \sigma(\mathbf{R})$, the operator $(\mathbf{R} - \lambda\mathbf{I})^{-1}$ is bounded. For $\mathbf{v} \in H \times H$. The solution $\mathbf{u}^*$ of the linear equation

$$(\mathbf{R}(\gamma) - \lambda\mathbf{I})\mathbf{u} = \mathbf{v}$$

is given by

$$\mathbf{u}^* = \mathbf{X}^{-1}(\alpha)(\mathbf{R} - \lambda\mathbf{I})^{-1}\mathbf{X}(\alpha)\mathbf{v}.$$

Since $\mathbf{X}(\gamma)$, $(\mathbf{R} - \lambda\mathbf{I})^{-1}$ and $X^{-1}(\gamma)$ are bounded operators, $\mathbf{u}^*$ verifies

$$\|\|\mathbf{u}^*\|\| \leq C\|\|\mathbf{v}\|\|,$$

where C is a constant independent of $\mathbf{v}$. It follows that $\lambda \notin \sigma(\mathbf{R}(\gamma))$.

Conversely, if $\lambda \notin \sigma(\mathbf{R}(\gamma))$, the operator $(\mathbf{R}(\gamma - \lambda I)^{-1}$ is bounded. For $\mathbf{v} \in H \times H$. The solution $\mathbf{u}^*$ of the linear equation

$$(\mathbf{R} - \lambda\mathbf{I})\mathbf{u} = \mathbf{v}$$

is given by

$$\mathbf{u}^* = \mathbf{X}^{-1}(\alpha)(\mathbf{R}(\gamma) - \lambda\mathbf{I})^{-1}\mathbf{X}(\alpha)\mathbf{v};$$

and verifies

$$\|\|\mathbf{u}^*\|\| \leq C\|\|\mathbf{v}\|\|,$$

where C is a constant independent of $\mathbf{v}$. It follows that $\lambda \notin \sigma(\mathbf{R})$. $\square$

Theorem 3.3.6. *Assume that $\omega \neq 0$. If $\lambda \in \sigma(\mathbf{L}_\omega(\alpha)) \setminus \{0\}$ and if τ satisfies*

$$(\lambda + \omega - 1)^2 = \lambda\omega^2\tau^2, \tag{50}$$

then $\tau \in \sigma(\mathbf{J}(\alpha))$. Conversely, if $\tau \in \sigma(\mathbf{J}(\alpha))$ and if λ satisfies (50), then $\lambda \in \sigma(\mathbf{L}_\omega(\alpha))$.

Proof. For $\omega \neq 0$, we have to prove that:

62

1. If $\lambda \notin \sigma(\mathbf{L}_\omega(\alpha))$ and if τ satisfies (50), then $\tau \notin \sigma(J(\alpha))$;

2. If then $\tau \notin \sigma(\mathbf{J}(\alpha))$ and if λ satisfies (50), then $\lambda \notin \sigma(\mathbf{L}_\omega(\alpha))$.

The operators $\mathbf{J}(\alpha)$ and $\mathbf{L}_\omega(\alpha)$ can be written as

$$\mathbf{L}_\omega(\alpha) = (\mathbf{I} + \omega \mathbf{R}_1(\alpha))^{-1} ((1-\omega)\mathbf{I} - \omega \mathbf{R}_2(\alpha))$$

$$\mathbf{J}(\alpha) = -(\mathbf{R}_1(\alpha) + \mathbf{R}_2(\alpha)),$$

where $\mathbf{R}_1(\alpha) = \begin{pmatrix} 0 & 0 \\ -P_1(\alpha) & 0 \end{pmatrix}$ and $\mathbf{R}_2(\alpha) = \begin{pmatrix} 0 & -A_1(\alpha) \\ 0 & 0 \end{pmatrix}$.

Let $\gamma \neq 0$. It holds from Proposition 3.3.5 that

$$\sigma(\mathbf{J}(\alpha)) = \sigma(\mathbf{R}(\gamma, \alpha)),$$

where $\mathbf{R}(\gamma, \alpha) = -(\frac{1}{\gamma}\mathbf{R}_1 + \gamma \mathbf{R}_2)$.

We assume that $\lambda \notin \sigma(\mathbf{L}_\omega(\alpha))$. Since $(\mathbf{L}_\omega(\alpha) - \lambda \mathbf{I})^{-1}$ is bounded, $\forall \mathbf{v} \in H \times H$, the function $\mathbf{u}$ defined by

$$\mathbf{u} = \lambda^{\frac{1}{2}}\omega(\mathbf{L}_\omega(\alpha) - \lambda \mathbf{I})^{-1}(I + \omega \mathbf{R}_1(\alpha))^{-1}\mathbf{v}, \tag{51}$$

satisfies

$$\|\|\mathbf{u}\|\| \leq C\|\|\mathbf{v}\|\|, \tag{52}$$

where C is a constant independent of $\mathbf{v}$.

Multiplying (51) by $(\mathbf{L}_\omega(\alpha) - \lambda \mathbf{I})$ and substituting $\mathbf{L}_\omega(\alpha)$ yields

$$(\mathbf{I} + \omega \mathbf{R}_1(\alpha))^{-1} ((1-\omega)\mathbf{I} - \omega \mathbf{R}_2(\alpha)) \mathbf{u} - \lambda \mathbf{u} = \lambda^{\frac{1}{2}}\omega(\mathbf{I} + \omega \mathbf{R}_1(\alpha))^{-1}\mathbf{v},$$

which is equivalent to

$$(-\omega(\mathbf{R}_2(\alpha) + \lambda \mathbf{R}_1(\alpha)) - (\lambda + \omega - 1)I) \mathbf{u} = \lambda^{\frac{1}{2}}\omega\mathbf{v}. \tag{53}$$

Multiplying (53) by $\lambda^{-\frac{1}{2}}\omega^{-1}$ yields

$$\left(\mathbf{R}(\lambda^{-\frac{1}{2}}, \alpha) - \frac{(\lambda + \omega - 1)}{\lambda^{\frac{1}{2}}\omega}\mathbf{I}\right) \mathbf{u} = \mathbf{v}. \tag{54}$$

It follows that (54) admits a solution given by (51) which satisfies (52). Thus $\tau = \frac{(\lambda+\omega-1)}{\lambda^{\frac{1}{2}}\omega} \notin \sigma(\mathbf{R}(\lambda^{-\frac{1}{2}}, \alpha)) = \sigma(\mathbf{J}(\alpha))$.

Conversely, assume that $\tau \notin \sigma(\mathbf{J}(\alpha)) = \sigma(\mathbf{R}(\lambda^{-\frac{1}{2}}, \alpha))$ and τ satisfies (50). $\forall \mathbf{v} \in H \times H$, the function u defined by

$$\mathbf{u} = (\lambda^{\frac{1}{2}}\omega)^{-1}(R(\lambda^{-\frac{1}{2}}, \alpha) - \tau I)^{-1}(I + \omega R_1(\alpha))\mathbf{v}, \tag{55}$$

satisfies the inequality (52). From (55), it follows that

$$(-\omega(\mathbf{R}_2(\alpha) + \lambda \mathbf{R}_1(\alpha)) - (\lambda + \omega - 1)\mathbf{I})\mathbf{u} = (\mathbf{I} + \omega \mathbf{R}_1(\alpha))\mathbf{v},$$

or equivalently

$$((1 - \omega)\mathbf{I} - \omega \mathbf{R}_2(\alpha))\mathbf{u} - \lambda(\mathbf{I} + \omega \mathbf{R}_1(\alpha))\mathbf{u} = (\mathbf{I} + \omega \mathbf{R}_1(\alpha))\mathbf{v}. \tag{56}$$

Since $(\mathbf{I} + \omega \mathbf{R}_1(\alpha))$ is invertible, (56) is equivalent to

$$(\mathbf{I} + \omega \mathbf{R}_1(\alpha))^{-1}((1 - \omega)\mathbf{I} - \omega \mathbf{R}_2(\alpha))\mathbf{u} - \lambda \mathbf{u} = \mathbf{v}. \tag{57}$$

It follows that $\forall \mathbf{v} \in H \times H$, the problem

$$(\mathbf{L}_\omega(\alpha) - \lambda \mathbf{I})\mathbf{u} = \mathbf{v}$$

has a solution $\mathbf{u}$ satisfying (52). Thus $\lambda \notin \sigma(\mathbf{L}_\omega(\alpha))$. $\qquad \square$

Remark 3.3.2. *The relation (50) between the spectrums of the operators of the Jacobi and SOR methods is the same as in the presence of finite dimensional linear systems with coefficient matrices possessing block Property A. Therefore, the results obtained in the last case may be generalized for the SOR acceleration of the PAS method.*

In particular, we have theoretically that Jacobi and Gauss-Seidel methods converge simultaneously and the Gauss-Seidel method is two times faster than Jacobi method. We also have the following convergence results of the SOR method (see [21, 50, 89]).

Theorem 3.3.7. *Convergence of the SOR method.*

1. *If $\sigma(\mathbf{J}(\alpha) \subset \mathbb{R}$, the SOR method converges for $0 < \omega < 2$ and the optimal convergence parameter is*

$$\omega_{opt} = \frac{2}{1 + \sqrt{1 - (r(\mathbf{J}(\alpha)))^2}}, \tag{58}$$

   where $r(\mathbf{J}(\alpha))$ is the spectral radius of $\mathbf{J}(\alpha)$.

64

2. *If $\sigma(\mathbf{J}(\alpha))$ contains complex numbers, the SOR method converges if for some positive number $\tau \in (0,1)$ and each $\lambda = \mu + i\beta \in \sigma(\mathbf{J}(\alpha))$, the point (μ, β) lies in the interior of the ellipse $\epsilon(1,\tau) = \left\{ (\mu, \beta) : \mu^2 + \frac{\beta^2}{\tau^2} = 1 \right\}$ and $0 < \omega < \frac{2}{1+\tau}$. Theoretical results for the determination of the SOR optimal parameter are given in [89].*

3.3.3 Numerical Results

We apply the PAS method on an example problem of particle transport.

Let $D = X \times W$, with $X = (0,1)$ and $W = [-1,1]$. We consider in $L^2(D)$ the following integro-differential equation:

$$\begin{cases} \mu \frac{\partial u}{\partial x} + \sigma(x)u = \int_{-1}^{1} \kappa(r,\mu,\mu')u(r,\mu')d\mu' + q(r,\mu) \ \text{ in } D \\ u(0,\mu) = 0, \ \mu > 0 \ \text{ and } \ u(1,\mu) = 0, \ \mu < 0 \end{cases}, \tag{59}$$

where function $\sigma(x)$ and $\kappa(x,\mu,\mu')$ satisfy the following conditions:

1. $\sigma \in L^\infty((0,1))$, $\exists \sigma_0 > 0$ such that $\sigma(x) \geq \sigma_0$ a.e. on $(0,1)$;

2. $\kappa(r,\mu,\mu') > 0$ a.e. on D and $\exists c \in (0,1)$, $\int_{-1}^{1} \kappa(r,\mu,\mu')d\mu' \leq \sigma_0 c$ a.e. on W.

In the operator form, equation (59) reads:

$$Tu = Au + Pu = q, \tag{60}$$

where the

$$Au = \mu \frac{\partial u}{\partial x} \ \text{ and } \ Pu = \sigma(x)u - \int_{-1}^{1} \kappa(r,\mu,\mu')u(r,\mu')d\mu'. \tag{61}$$

The operator A is m-accretive in $L^2(D)$ and it follows from the assumption made on σ and κ that operator P is positive definite [34]. The PAS iteration method can be applied for the solution of equation (59). The equation (59) is known to be near singular when $c \approx 1$ [34].

The discretization is carried out by a DSN scheme [34] consisting of using a finite set of L discrete angular directions $\{\mu_k\}_{l=1}^{L} \in [-1,1]$, which are nonzero and symmetric about the origin for the angular approximation and a difference method based on control volume approach and cell averaging for the spatial approximation.

For the numerical results, we took particular data for which an exact solution u is known: $\sigma(x) = \sigma$, $\kappa(x,\mu,\mu') = \frac{3\sigma c}{2}\mu'^2$, $(0 < c < 1)$,

$$q(x,\mu) = \begin{cases} \mu^2 + \sigma\mu x - \frac{3\sigma c}{8} & \mu > 0 \\ \mu^2 + \sigma\mu(x-1) - \frac{3\sigma c}{8} & \mu < 0 \end{cases} ; \ u(x,\mu) = \begin{cases} \mu x & \mu > 0 \\ \mu(x-1) & \mu < 0 \end{cases}.$$

65

We study the behavior of PAS, Jacobi (Jacobi_PAS), Gauss-Seidel (GS_PAS) and SOR (SOR_PAS) algorithms with respect to parameters α, c and σ. For iterative methods tested here, the iterations are stopped when the relative error $\frac{\|U_{exact} - U^{(k)}\|_2}{\|U_{exact}\|_2}$ is less than a prescribed $\epsilon > 0$. The spatial mesh size is $h = 1/30$ and the angular mesh size is $\tau = 1/500$.

There are three sets of tests: one for fixed c, another for fixed σ and the last for fixed α. For fixed σ or c, we set $\alpha = \sigma(1-c)$. As shown by Figure 5 to Figure 8 all the methods converge. For $\sigma = 100$, we compare the c-dependence of the iterative methods used here (Figure 5 and Figure 6). We observe that Jacobi method is slower than PAS method which has the same convergence behavior as the Gauss-Seidel method except for the values of $c > 0.98$ (Figure 6), where the convergence of PAS seems to slow down. The σ-dependence of the iterative methods at fixed $c = 0.8$ is plotted by Figure 7 and Figure 8. We notice the convergence of all methods even for very large values of σ. The Jacobi method remains slower than the PAS method which behaves as Gauss-Seidel method. It can be observed in the two sets of test that SOR is the fastest method. We also remark that, Gauss-Seidel is two times faster than Jacobi. This confirms the theoretical convergence results obtained. At fixed $\alpha = 5$, we compare the σ-dependence at $c = 0.5$(Figure 9) and the c-dependence at $\sigma = 100$ (Figure 10) of the PAS and SOR methods. The SOR method gives excellent results than PAS. Figure 11 plots the convergence behavior of PAS and SOR methods at fixed $\alpha = 10$ for $c \in \{0.5, 0.9\}$ and $\sigma \in \{100, 1000\}$.

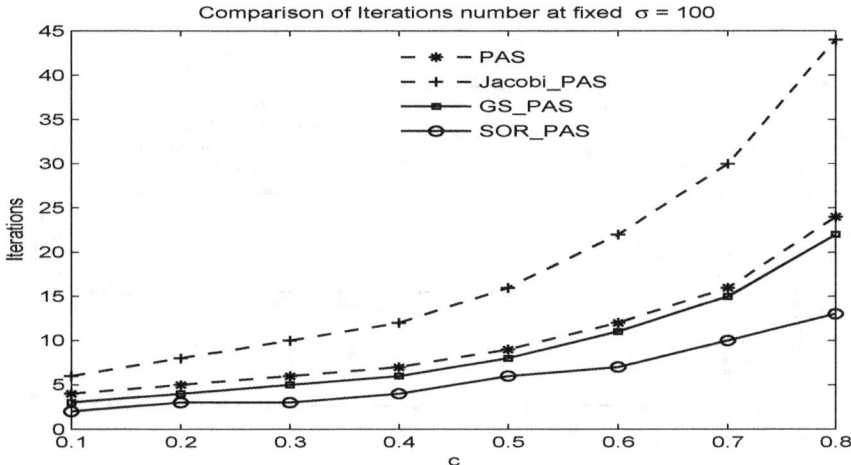

Figure 5: Comparison of the PAS, Jacobi, Gauss-Seidel and SOR ($\omega = 0.95$) methods at fixed $\sigma = 100$, for $c \in [0.1, 0.8](\epsilon = 5E - 05)$.

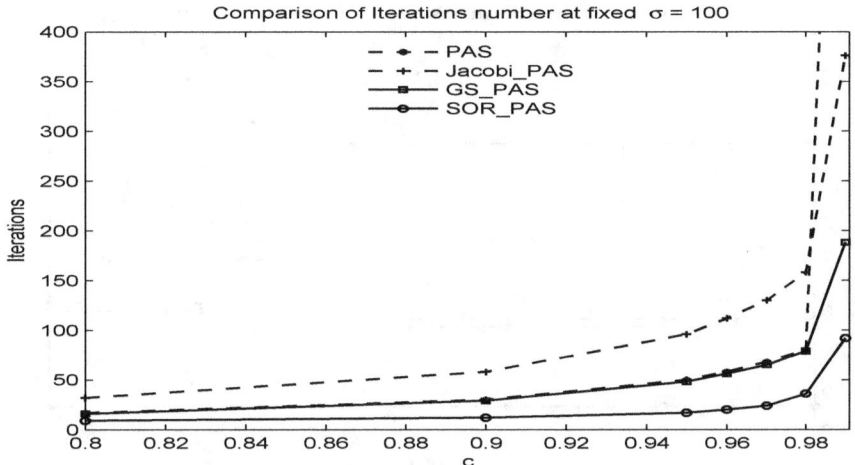

Figure 6: Comparison of the PAS, Jacobi, Gauss-Seidel and SOR ($\omega = 0.95$) methods at fixed $\sigma = 100$, for $c \in [0.8, 0.99]$ ($\epsilon = 5E - 04$).

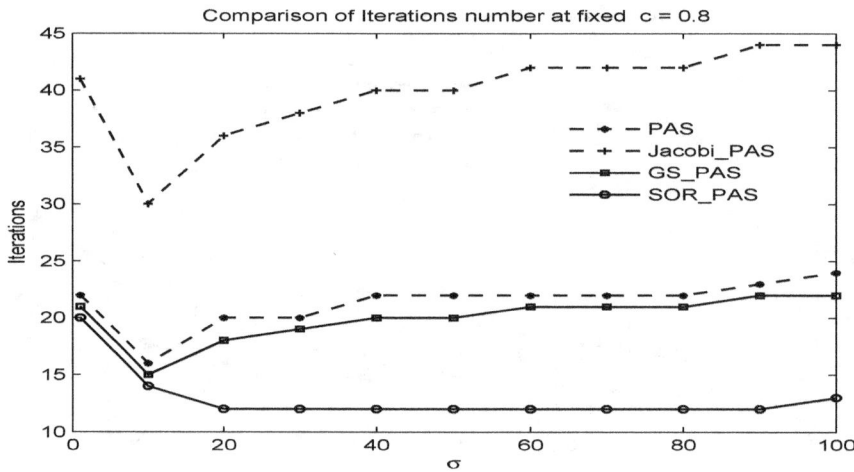

Figure 7: Comparison of the PAS, Jacobi, Gauss-Seidel and SOR ($\omega = 0.95$) methods at fixed $c = 0.8$, for $\sigma \in [1, 100]$ ($\epsilon = 5E - 05$).

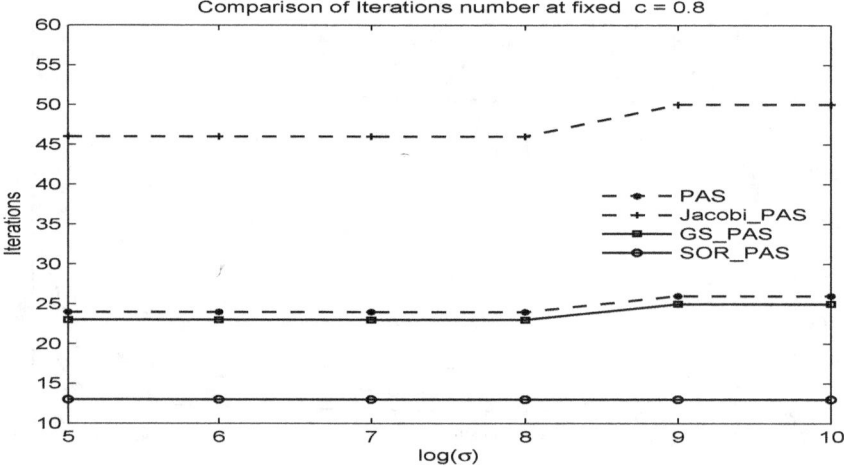

Figure 8: Comparison of the PAS, Jacobi, Gauss-Seidel and SOR ($\omega = 0.95$) methods at fixed $c = 0.8$, for large values of σ ($\epsilon = 5E - 05$).

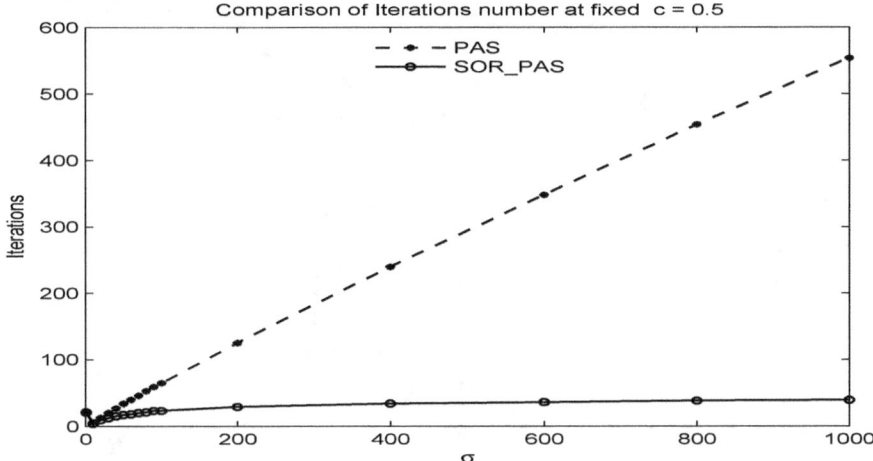

Figure 9: Iteration number as function of σ for PAS and SOR ($\omega = 0.95$) methods at fixed $c = 0.8$ and $\alpha = 5$ ($\epsilon = 5E - 04$).

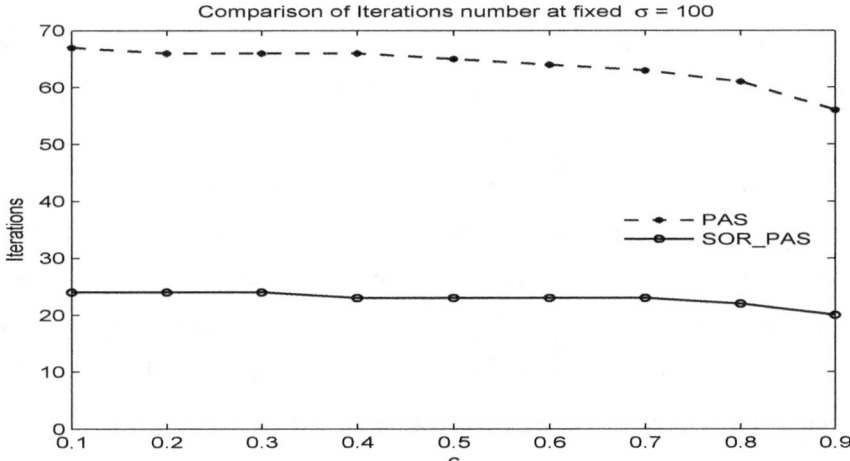

Figure 10: Iteration number as function of c for PAS and SOR ($\omega = 0.95$) methods at fixed $c\sigma = 100$ and $\alpha = 5$ ($\epsilon = 5E - 04$).

69

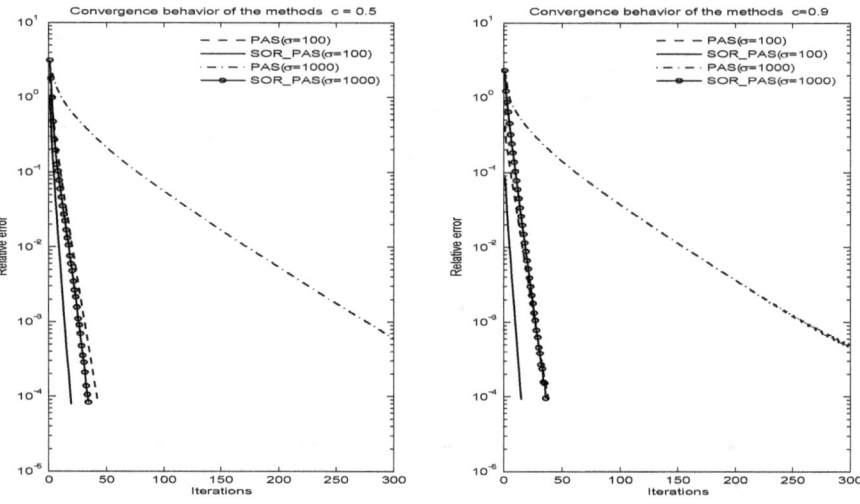

Figure 11: Convergence behavior of PAS and SOR ($\omega = 0.95$) methods at fixed $\alpha = 10$.

3.4 Self Adjoint and m-Accretive Splitting Iterative Method

We consider a particular case of the PAS iteration method introduced in section 3.3, where the positive definite part of the operator is Self-adjoint. In this work, focus is given on iterative methods for the numerical treatment of the single group steady state neutron transport equation in slab geometry, in 1D-spherical geometry and bounded convex domain of $\mathbb{R}^n$ ($n = 2, 3$). We introduce a two-step iteration method linked to a Self-adjoint and m-Accretive splitting of the transport operator. We investigate the convergence of this iterative method. Theoretical analysis shows that this iterative method converges unconditionally to the solution of the transport equation. An upper bound for the contraction factor of the iteration is derived. It is solely dependent on the spectrum of the Self-adjoint part of the transport operator. Moreover, the convergence analysis of the incomplete version of the SAS iteration is provided. The convergence of the method is numerically illustrated and compare with the standard Source Iteration method, a spatial multigrid method and Krylov subspace methods such as Induced Dimension Reduction (IDR) (see [76]), BiCGStab (see [31, 32]) and a preconditioned GMRES methods (see

[22]) on sample problems in slab geometry, in spherical geometry and in two dimensional space.

The remainder of this section is organized as follows. In subsection 3.4.1, we present the splitting of the transport operator. The Subsection 3.4.2 is devoted the the presentation of the SAS iteration method. The Subsection 3.4.3 is dedicated to the convergence analysis of the iterative methods. Subsection 3.4.4 deals with an implementation of the method. Convergence results of the SOR acceleration of the SAS iteration are given in subsection 3.4.5. The discretization and the numerical results obtained from example problems in slab geometry, 1D-spherical geometry and two dimensional space are presented in subsection 3.4.6.

3.4.1 The Splitting Methods.

Otherwise specify, the notations and definitions of section 3.2 hold in the remainder of this section.

We consider the neutron transport problem (4)-(5), written as:

$$
\begin{cases}
T\psi(x, \Omega) = q(x, \Omega) \\
\psi \in W_0,
\end{cases}
\tag{62}
$$

where the operator $T = A + \Sigma - K$, with operators A, Σ and K given by (10). The functional space W_0 is defined by (8). We have $\mathcal{D}(T) = W_0$.

The standard splitting of the transport operator consists of a decoupling between the differential and the integral parts as follows

$$
-T = K - L,
\tag{63}
$$

where $L = A + \Sigma$. This splitting leads to the source iteration method defined by: given $\psi_0 \in \mathcal{D}(T)$, solve

$$
\begin{cases}
L\psi^{(n+1)} = K\psi^n + q \text{ in } \Omega \\
\psi^{(n+1)} \in \mathcal{D}(T)
\end{cases}.
\tag{64}
$$

This method becomes extremely slow in the critical case. Several acceleration techniques of the convergence of the source iteration method such as Diffusion Synthetic Acceleration (DSA) (see [84]) and multigrid algorithms have been introduced and studied (see [30, 56]).

Let us consider now another natural splitting of the transport operator stated as follows:

$$
-T = -(A + S),
\tag{65}
$$

71

where $S = \Sigma - K$. Therefore for any positive constant α, we have the following two-step splitting:

$$
\begin{cases}
T = (\alpha I + S) - (\alpha I - A) \\
T = (\alpha I + A) - (\alpha I - S)
\end{cases}
\tag{66}
$$

3.4.2 The SA Splitting Iteration.

We present in this section an iterative methods for the solution of the problem (4)-(5). The method relies on the splitting (66), is a two-step iteration method.

Let's consider the constant $\alpha > 0$. Given an initial guess $\psi^{(0)} \in D(T)$, for $k = 0, 1, \ldots$ until $\{\psi^{(k)}\}$ converges, calculate

$$
\begin{cases}
(\alpha I + S)\psi^{(k+\frac{1}{2})} = (\alpha I - A)\psi^{(k)} + q \\
(\alpha I + A)\psi^{(k+1)} = (\alpha I - S)\psi^{(k+\frac{1}{2})} + q
\end{cases}
\tag{67}
$$

Remark 3.4.1. *We have the following observations:*

1. *Since the operator S is bounded, self-adjoint and positive definite, the operator $\alpha I + S$ is bounded self-adjoint and positive definite for $\alpha > 0$. It then follows that $(\alpha I + S)^{-1}$ exists, is bounded and self-adjoint. We also have that $\alpha I - S$ is a bounded self-adjoint operator.*

2. *From the m-accretive property of the operator A, we deduce that for $\alpha > 0$, the operator $\alpha I + A$ is invertible from $D(T)$ to $L^2(Q)$ and its inverse $(\alpha I + A)^{-1}$ is a bounded operator.*

From equation (67), we deduce that $\psi^{(k+1)}$ satisfies

$$
(\alpha I + A)\psi^{(k+1)} = M(\alpha)(\alpha I + A)\psi^{(k)} + N(\alpha)q,
\tag{68}
$$

where

$$
M(\alpha) = S(\alpha)A(\alpha) \text{ and } N(\alpha) = 2\alpha(\alpha I + S)^{-1};
\tag{69}
$$

with $S(\alpha) = (\alpha I - S)(\alpha I + S)^{-1}$ and $A(\alpha) = (\alpha I - A)(\alpha I + A)^{-1}$.

Therefore, the exact solution ψ^* of the problem (4)-(5) verifies

$$
\|\psi^{(k+1)} - \psi^*\|_{A(\alpha)} \leq \|M(\alpha)\| \|\psi^{(k)} - \psi^*\|_{A(\alpha)},
\tag{70}
$$

where $\|\psi\|_{A(\alpha)} = \|(\alpha I + A)\psi\|$, $\psi \in W_0$.

Let α be a positive constant. Since the functional $\|\|.\|\|_{A(\alpha)}$ is a norm on W_0 equivalent to $\|.\|_{W^2}$ (see Theorem 3.2.4), it is well known that the iteration method (67) converges if the operator $M(\alpha)$ satisfies

$$\|(M(\alpha)\| < 1. \tag{71}$$

Theorem 3.4.1. *Convergence of the SAS iteration method.*

Let α be a positive constant. The norm $\|(M(\alpha)\|$ of the operator $M(\alpha)$ is bounded by

$$\beta(\alpha) = \sup_{\lambda \in \sigma(S)} \left| \frac{\alpha - \lambda}{\alpha + \lambda} \right|, \tag{72}$$

where $\sigma(S)$ is the spectrum of the operator S. Therefore it holds that

$$\|(M(\alpha)\| \leq \beta(\alpha) < 1, \ \forall \alpha > 0, \tag{73}$$

and the SAS iteration converges to the unique solution $\psi^* \in \mathcal{D}(T)$ of the problem (4)-(5). The optimal parameter $\bar{\alpha}$ which minimizes the bound $\beta(\alpha)$ is given by [59]

$$\bar{\alpha} = \sqrt{\lambda_{\min}\lambda_{\max}} \tag{74}$$

and

$$\beta(\bar{\alpha}) = \frac{\sqrt{\lambda_{\max}} - \sqrt{\lambda_{\min}}}{\sqrt{\lambda_{\max}} + \sqrt{\lambda_{\min}}}, \tag{75}$$

where $\lambda_{\min}$ and $\lambda_{\max}$ denote respectively the lower and the upper bounds of the spectrum of the operator S.

Proof. For $\alpha > 0$, $A(\alpha)$ and $S(\alpha)$ are bounded operators from $L^2(Q)$ to $\mathcal{D}(T)$ and $L^2(Q)$ respectively. We have

$$\|M(\alpha)\| \leq \|S(\alpha)\| \, \|A(\alpha)\|.$$

Since S is a bounded self-adjoint operator in $L^2(Q)$, we have

$$\|S(\alpha)\| = \|(\alpha I - S)(\alpha I + S)^{-1}\| = \sup_{\lambda \in \sigma(S)} \left| \frac{\alpha - \lambda}{\alpha + \lambda} \right| = \beta(\alpha).$$

It holds from the positivity of the reals α and λ that $\beta(\alpha) < 1$.

The proof of the theorem is achieved if the norm of operator $A(\alpha)$ verifies $\|A(\alpha)\| \leq 1$.

The operator A is m-accretive in the Hilbert space $L^2(Q)$. It follows from Theorem 3.2.3 that $\|A(\alpha)\| \leq 1$. Therefore, $\|M(\alpha)\| \leq \beta(\alpha) < 1$. $\qquad \square$

From the equivalence between the norms $\|.\|_{W^2}$ and $\|.\|_{(A(\alpha)}$ follows the convergence with respect to the norm $\|.\|_{W^2}$ in $\mathcal{D}(T)$. Additionally, since

$$\|u^{(k+1)} - u^*\| \leq \|u^{(k+1)} - u^*\|_{W^2}; \;\; k = 0, 1, 2, \cdots, \tag{76}$$

we have

$$\lim_{k \to +\infty} \|u^{(k)} - u^*\| = \lim_{k \to +\infty} \|u^{(k)} - u^*\|_{W^2} = 0. \tag{77}$$

Thus the sequence $u^{(k)}$ converges in $\mathcal{D}(T)$ with respect to the norm $\|.\|$.

The results of Theorem 3.4.1 show that $\beta(\alpha)$ is an upper bound of the contraction factor of the SAS iteration in the sense of the norm $\|.\|_{(A+\alpha I)}$ in $\mathcal{D}(T)$.

It follows from assumption (A1), (A2) and (A3) that (see [34, 61]): $\|K\| \leq \sigma_0 c$ and for any $u \in L^2(Q)$, $(Su, u) \geq \sigma_0(1 - c)\|u\|^2$. Since S is self-adjoint, bounded and positive definite, S^{-1} is self-adjoint bounded and positive definite on the Hilbert space $L^2(Q)$. It holds that (see [24]):

$$\frac{1}{\lambda_{\max}} \leq \|S^{-1}\| = \sup_{\|u\| \neq 0} \frac{\|u\|}{\|Su\|} \leq \frac{1}{\sigma_0(1 - c)}.$$

It follows that $\lambda_{\min} \geq \sigma_0(1 - c)$ and

$$\sigma(S) \subset [\sigma_0(1 - c), \sigma_1 + \sigma_0 c], \tag{78}$$

where $\sigma_1 = \sup\limits_{x \in D} |\sigma(x)|$. We then have

$$\beta(\alpha) \leq \sup_{\lambda \in [\sigma_0(1-c), \sigma_1 + \sigma_0 c]} \left| \frac{\lambda - \alpha}{\lambda + \alpha} \right| < 1. \tag{79}$$

Each step of the SAS iterative method is constituted of two-half steps which require finding solutions of linear equations with operators $(\alpha I + S)$ and $(\alpha I + A)$. Exact solutions of these equations are generally not available. These linear equations can be solve approximately using appropriate methods with respect to the properties of each operators. This results in the following inexact Self-adjoint/m-Accretive splitting (ISAS) iteration for solving the linear equation (4)-(5).

3.4.3   The ISAS Iteration Method.

Given an initial guess $\bar{\psi}^{(0)} \in \mathcal{D}(T)$. For $k = 0, 1, 2, \ldots$ until $\{\bar{\psi}^{(k)}\}$ converges, solve $\bar{\psi}^{(k+\frac{1}{2})}$ approximately from

$$(\alpha I + S)\bar{\psi}^{(k+\frac{1}{2})} \approx (\alpha I - A)\bar{\psi}^{(k)} + q \tag{80}$$

74

by employing an inner iteration (e.g the conjugate gradient method) with $\bar{\psi}^{(k)}$ as the initial guess, then solve $\bar{\psi}^{(k+1)}$ approximately from

$$(\alpha I + A)\bar{\psi}^{(k+1)} \approx (\alpha I - S)\bar{\psi}^{(k+\frac{1}{2})} + q \tag{81}$$

where α is a given positive constant.

We now set

$$M_1 = \alpha I + S, \; M_2 = \alpha I + A, \; N_1 = \alpha I - A \text{ and } N_2 = \alpha I - S. \tag{82}$$

Theorem 3.4.2. *Convergence of the ISAS iteration method.*

If the iterative sequence $\{\bar{\psi}^{(k)}\}$ is defined as follows

$$\bar{\psi}^{(k+\frac{1}{2})} = \bar{\psi}^{(k)} + \bar{\varphi}^{(k)}, \; \text{with } M_1\bar{\varphi}^{(k)} = \bar{\phi}^{(k)} + \bar{p}^{(k)}, \tag{83}$$

satisfying $\frac{\left\|\bar{p}^{(k)}\right\|}{\left\|\bar{\phi}^{(k)}\right\|} \leq \epsilon_k$, where $\bar{\phi}^{(k)} = q - T\bar{\psi}^{(k)}$; and

$$\bar{\psi}^{(k+1)} = \bar{\psi}^{(k+\frac{1}{2})} + \bar{\varphi}^{(k+\frac{1}{2})}, \; \text{with } M_2\bar{\varphi}^{(k+\frac{1}{2})} = \bar{\phi}^{(k+\frac{1}{2})} + \bar{\tau}^{(k+\frac{1}{2})}, \tag{84}$$

satisfying $\frac{\left\|\bar{\tau}^{(k+\frac{1}{2})}\right\|}{\left\|\bar{\phi}^{(k)}\right\|} \leq \eta_k$, where $\bar{\phi}^{(k+\frac{1}{2})} = q - T\bar{\psi}^{(k+\frac{1}{2})}$, then $\{\bar{\psi}^{(k)}\}$ is of the form

$$\begin{aligned}
\bar{\psi}^{(k+1)} &= M_2^{-1}N_2M_1^{-1}N_1\bar{\psi}^{(k)} + M_2^{-1}(I + N_2M_1^{-1})q \\
&\quad + M_2^{-1}(N_2M_1^{-1}\bar{p}^{(k)} + \bar{\tau}^{(k+\frac{1}{2})}).
\end{aligned} \tag{85}$$

Moreover, if $\psi^* \in \mathcal{D}(T)$ is the exact solution of the equation (4)-(5), then we have for $k = 0, 1, 2, \ldots$:

$$\left\|M_2(\bar{\psi}^{(k+1)} - \psi^*)\right\| \leq (\beta + \mu\theta(\mu\epsilon_k + \eta_k))\left\|M_2(\bar{\psi}^{(k)} - \psi^*)\right\|, \tag{86}$$

where

$$\beta = \left\|N_2M_1^{-1}N_1M_2^{-1}\right\| \leq \beta(\alpha), \quad \mu = \left\|N_2M_1^{-1}\right\|, \quad \theta = \left\|TM_2^{-1}\right\|. \tag{87}$$

In particular, if

$$\beta(\alpha) + \theta(\mu\epsilon_{\max} + \eta_{\max}) < 1, \tag{88}$$

then the iterative sequence $\{\bar{\psi}^{(k)}\}$ converges to $\psi^* \in \mathcal{D}(T)$, where $\epsilon_{\max} = \max\limits_k\{\epsilon_k\}$ and $\eta_{\max} = \max\limits_k\{\eta_k\}$.

Proof. From (83), we have

$$\bar{\psi}^{(k+\frac{1}{2})} = \bar{\psi}^{(k)} + M_1^{-1}(\bar{\phi}^{(k)} + \bar{p}^{(k)})$$
$$= (I - M_1^{-1}T)\bar{\psi}^{(k)} + M_1^{-1}q + M_1^{-1}\bar{p}^{(k)}$$
$$= M_1^{-1}N_1\bar{\psi}^{(k)} + M_1^{-1}q + M_1^{-1}\bar{p}^{(k)}. \tag{89}$$

Similarly, from (84) we have

$$\bar{\psi}^{(k+1)} = M_2^{-1}N_2\bar{\psi}^{(k+\frac{1}{2})} + M_2^{-1}q + M_2^{-1}\bar{\tau}^{(k+\frac{1}{2})}$$
$$= M_2^{-1}N_2(M_1^{-1}N_1\bar{\psi}^{(k)} + M_1^{-1}q + M_1^{-1}\bar{p}^{(k)}) + M_2^{-1}q + M_2^{-1}\bar{\tau}^{(k+\frac{1}{2})}$$
$$= M_2^{-1}N_2M_1^{-1}N_1\bar{\psi}^{(k)} + M_2^{-1}(N_2M_1^{-1} + I)q$$
$$+ M_2^{-1}(N_2M_1^{-1}\bar{p}^{(k)} + \bar{\tau}^{(k+\frac{1}{2})}). \tag{90}$$

The exact solution ψ^* of equation (4)-(5) satisfies

$$\psi^* = M_2^{-1}N_2M_1^{-1}N_1\psi^* + M_2^{-1}(N_2M_1^{-1} + I)q. \tag{91}$$

By subtracting (91) from (90), we have

$$\bar{\psi}^{(k+1)} - \psi^* = M_2^{-1}N_2M_1^{-1}N_1(\bar{\psi}^{(k)} - \psi^*) + M_2^{-1}(N_2M_1^{-1}\bar{p}^{(k)} + \bar{\tau}^{(k+\frac{1}{2})}) \tag{92}$$

Multiplying (92) by M_2 and taking norms on both sides, we obtain

$$\left\| M_2(\bar{\psi}^{(k+1)} - \psi^*) \right\| \leq \left\| N_2M_1^{-1}N_1M_2^{-1}M_2(\bar{\psi}^{(k)} - \psi^*) \right\|$$
$$+ \left\| N_2M_1^{-1} \right\| \left\| \bar{p}^{(k)} \right\| + \left\| \bar{\tau}^{(k+\frac{1}{2})} \right\|$$
$$\leq \left\| N_2M_1^{-1}N_1M_2^{-1} \right\| \left\| M_2(\bar{\psi}^{(k)} - \psi^*) \right\|$$
$$+ \left\| N_2M_1^{-1} \right\| \left\| \bar{p}^{(k)} \right\| + \left\| \bar{\tau}^{(k+\frac{1}{2})} \right\|.$$

Noticing that

$$\left\| \bar{\phi}^{(k)} \right\| = \left\| b - T\bar{\psi}^{(k)} \right\| = \left\| T(\psi^* - \bar{\psi}^{(k)}) \right\| \leq \left\| TM_2^{-1} \right\| \left\| M_2(\psi^* - \bar{\psi}^{(k)}) \right\|,$$

by the definition of the sequences $\{\bar{p}^{(k)}\}$ and $\{\tau^{(k+\frac{1}{2})}\}$ we have

$$\left\| \bar{p}^{(k)} \right\| \leq \epsilon_k \left\| \bar{\phi}^{(k)}) \right\| \leq \epsilon_k \left\| TM_2^{-1} \right\| \left\| M_2(\psi^* - \bar{\psi}^{(k)}) \right\|$$

and

$$\left\| \bar{\tau}^{(k+\frac{1}{2})} \right\| \leq \eta_k \left\| \bar{\phi}^{(k)} \right\| \leq \eta_k \left\| T M_2^{-1} \right\| \left\| M_2(\psi^* - \bar{\psi}^{(k)}) \right\|.$$

through substituting, we finally obtain

$$\begin{aligned}
\left\| M_2(\bar{\psi}^{(k+1)} - \psi^*) \right\| &\leq \left\| N_2 M_1^{-1} N_1 M_2^{-1} \right\| \left\| M_2(\bar{\psi}^{(k)} - \psi^*) \right\| \\
&+ \epsilon_k \left\| N_2 M_1^{-1} \right\| \left\| T M_2^{-1} \right\| \left\| M_2(\psi^* - \bar{\psi}^{(k)}) \right\| \\
&+ \eta_k \left\| T M_2^{-1} \right\| \left\| M_2(\bar{\psi}^{(k)} - \bar{\psi}^*) \right\| \\
&\leq (\beta + \theta(\mu\epsilon_k + \eta_k)) \left\| M_2(\bar{\psi}^{(k)} - \psi^*) \right\|.
\end{aligned}$$

$\square$

Remark 3.4.2. *If the first (Resp. the second) inner equation of 67 can be solved exactly, then the sequence ϵ_k (Resp. η_k) is equal to zero and the convergence rate of the ISAS iteration is given by:*

$$R = \beta(\alpha) + \theta\eta_{\max} \ (Resp. \ R = \beta(\alpha) + \mu\theta\epsilon_{\max}). \tag{93}$$

It then follows that the convergence rate of the ISAS iterations is reduced to the same as that of SAS iteration when the two inner equations of (67) are exactly solved.

Theorem 3.4.3. *In the case of isotropic scattering where the integral operator is defined by*

$$K\psi = \sigma(x)cP\psi, \tag{94}$$

with

$$P\psi = \frac{1}{|S_2|} \int_{S_2} \psi(x, \boldsymbol{\Omega}')d\boldsymbol{\Omega}', \ |S_2| = \int_{S_2} d\boldsymbol{\Omega}';$$

the inverse of the operator $(\alpha I + S)$ is given by:

$$(\alpha I + S)^{-1} = \frac{1}{\sigma(x)(1-c) + \alpha} P + \frac{1}{\sigma(x) + \alpha}(I - P). \tag{95}$$

Proof. The operator $(\alpha I + S)$ can be written as $(\alpha I + S) = \lambda(P + \epsilon(I - P))$, where $\lambda = \sigma(x)(1-c) + \alpha$ and $\epsilon = \frac{\sigma(x)+\alpha}{\lambda}$.

Since $P^2 = P$, one has $(P + \epsilon(I-P))\left(P + \frac{1}{\epsilon}(I-P)\right) = I$. It then follows that $(\alpha I + S)^{-1} = \frac{1}{\lambda}\left(P + \frac{1}{\epsilon}(I-P)\right).$ $\square$

Therefore, in the case of isotropic scattering, the first subproblem of the system (67) can be solved explicitly. Moreover, the two-step iteration method defined by (67) can be reduced to the following iteration: Given $\psi^{(0)} \in W_0$,

$$(\alpha I + A)\psi^{(k+1)} = B\left[(\alpha I - A)\psi^{(k)} + q\right] + q, \text{ for } k = 0, 1, ... \tag{96}$$

where

$$B = \left(\frac{\alpha - \sigma(1-c)}{\alpha + \sigma(1-c)} - \frac{\alpha - \sigma}{\alpha + \sigma}\right) P + \frac{\alpha - \sigma}{\alpha + \sigma} I. \tag{97}$$

3.4.4 An Implementation of the ISAS Method.

The SAS iteration (67), can be written as : Given an initial guess $\varphi^{(0)} \in L^2(\Omega)$, for $k = 0, 1, \ldots$ until $\{\varphi^{(k)}\}$ converges, calculate

$$\begin{cases} (\alpha I + S)\varphi^{(k+\frac{1}{2})} = (\alpha I - A)(\alpha I + A)^{-1}\varphi^{(k)} + q \\ \varphi^{(k+1)} = (\alpha I - S)\varphi^{(k+\frac{1}{2})} + q \end{cases}, \tag{98}$$

where

$$\varphi^{(k)} = (\alpha I + A)\psi^{(k)} \quad \text{and} \quad \varphi^{(k+\frac{1}{2})} = \psi^{(k+\frac{1}{2})}. \tag{99}$$

At each step k of the iteration method (98), we have to solve a linear equation

$$\mathcal{A}(\alpha)F = q(\alpha), \tag{100}$$

where $\mathcal{A}(\alpha) = (\alpha I + S)$ and $q(\alpha) = (\alpha I - A)(\alpha I + A)^{-1}\varphi^{(k)} + q$. The solution of equation (100) is then used to compute $\varphi^{(k+1)}$. An infinite dimensional adaptation of the conjugate gradient method is employed to solve the equation (100). We have the following algorithm for the ISAS method:

Algorithm 3.4.1. *ISAS Algorithm*

Let $\psi^{(0)} \in \mathcal{D}(T)$, $\mathcal{R}^{(0)} = q - T\psi^{(0)}$, $q(\alpha) = (\alpha I - A)\psi^{(0)} + q$.

`While` $\|\mathcal{R}^{(k)}\| > \epsilon$ `do`

`begin`

    `solve` $\mathcal{A}(\alpha)F = q(\alpha)$ `by CG method;`

    `compute` $\varphi^{(k+1)} = (\alpha I - S)F + q$ `;`

    `compute` $q(\alpha) = (\alpha I - A)(\alpha I + A)^{-1}\varphi^{(k+1)}$ `;`

```
compute  R^(k+1) = q - (I - (αI - S)(αI + A)^{-1})φ^(k+1)  ;
```

`end.`

In the previous algorithms, we have to make clear how the right hand side $q(\alpha)$ is computed, since it contains the inverse operator $(\alpha I + A)^{-1}$. Let $\varphi \in L^2(Q)$, we describe in the following how to compute $\phi = (\alpha I - A)(\alpha I + A)^{-1}\varphi$. We have

$$\phi = (\alpha I - A)f, \tag{101}$$

where f verifies the linear equation

$$\begin{cases} (\alpha I + A)f = \varphi \\ f \in \mathcal{D}(T) \end{cases}. \tag{102}$$

Once f is calculated, the product (101) can be easily computed. The differential equation (102) can be solved numerically.

The solution of the discrete form of (102) can be approximated by preconditioned Krylov methods such as the BiCGStab algorithm with Gauss-Seidel preconditioner which was applied and investigated for simulation of convection dominated flows and heat conduction problems with nonlinear convection flows through boundaries of the domain (see [31, 32]). In the case of one and two dimensional spaces, where DSN schemes are used, the direct sweeping algorithm (see [34]) is advised since it solve the problem in $O(N)$ operations, N being the side of the problem.

3.4.5 SOR acceleration of the SAS Iteration Method

The following fixed point equation can be derived from the definition of the SAS iteration (67):

$$\mathbf{T}(\alpha)\mathbf{u} = \mathbf{q}, \tag{103}$$

where

$$\mathbf{T}(\alpha) = \begin{pmatrix} (\alpha I + S) & -(\alpha I - A) \\ -(\alpha I - S) & (\alpha I + A) \end{pmatrix}, \tag{104}$$

$$\mathbf{u} = \begin{pmatrix} u_1 \\ u_2 \end{pmatrix} \quad \text{and} \quad \mathbf{q} = \begin{pmatrix} q \\ q \end{pmatrix}. \tag{105}$$

Since A is m-accretive and S is positive definite, for any $\alpha > 0$, the solution of problem 103 exists an is unique in $\mathcal{D}(T) \times \mathcal{D}(T)$ (see Theorem 3.3.2). It also holds that $u^* \in \mathcal{D}(T)$ is solution to (62) if and only if $\mathbf{u}^* = \begin{pmatrix} u_{1*} \\ u_{2*} \end{pmatrix} \in \mathcal{D}(T) \times \mathcal{D}(T)$ is solution to (103) (see Theorem 3.3.3).

Let $\mathbf{u} = \begin{pmatrix} u_1 \\ u_2 \end{pmatrix} \in L^2(Q) \times L^2(Q)$. We consider in $L^2(Q) \times L^2(Q)$ the norm

$$|||\mathbf{u}||| = \left(\|u_1\|^2 + \|u_2\|^2 \right)^{\frac{1}{2}}. \tag{106}$$

Let consider in $\mathcal{D}(T) \times \mathcal{D}(T)$ the operator

$$\mathbf{P}(\alpha) = \begin{pmatrix} (\alpha I + S) & 0 \\ 0 & (\alpha I + A) \end{pmatrix}. \tag{107}$$

The preconditioning of the system (103) from the right by $[\mathbf{P}(\alpha)]^{-1}$ leads to the following system

$$\mathbf{T}_1(\alpha)\mathbf{u} = \mathbf{q} \tag{108}$$

$$\mathbf{T}_1(\alpha) = \begin{pmatrix} I & -A(\alpha) \\ -S(\alpha) & I \end{pmatrix}. \tag{109}$$

with $A(\alpha) = (\alpha I - A)(\alpha I + A)^{-1}$ and $S(\alpha) = (\alpha I - S)(\alpha I + S)^{-1}$. The operator $\mathbf{T}_1(\alpha)$ is bounded in $L^2(Q) \times L^2(Q)$ and the solution $\mathbf{v}^*$ (103) reads $\mathbf{v}^* = [\mathbf{P}(\alpha)]^{-1}\mathbf{u}^*$, where $\mathbf{u}^*$ is solution of (109).

The equation (108) can be solved using the Jacobi iteration method defined by (46) and the SOR iteration method given by (48), where the operators $A1(\alpha)$ and $P1(\alpha)$ have been replaced by $A(\alpha)$ and $S(\alpha)$ respectively.

Let $\mathbf{J}(\alpha)$ and $\mathbf{L}_\theta(\alpha)$ denote the Jacobi and SOR iteration operators respectively. It is well known that the choice of $\theta = 1$ results in the Gauss-Seidel iteration for solving the system (108).

Theorem 3.4.4. *Let α be a positive constant. It holds that:*

$$|||J(\alpha)||| \ < \ 1 \tag{110}$$

$$|||L_1(\alpha)||| \ < \ \beta(\alpha). \tag{111}$$

It the follows that the Jacobi and Gauss-Seidel iteration methods for the solution of (108) converge. Moreover, the Gauss-Seidel and SAS iterations are theoretically equivalent.

80

Proof. For $\mathbf{u} = \begin{pmatrix} u_1 \\ u_2 \end{pmatrix} \in L^2(\Omega) \times L^2(\Omega)$. Since A is m-accretive and S is positive definite, inequality (110) follows from Theorem 3.3.4.

Let $\mathbf{u}^* = \begin{pmatrix} u_{1*} \\ u_{2*} \end{pmatrix}$ be the exact solution to (108) and $\mathbf{u}^{(k)} = \begin{pmatrix} u_1^{(k)} \\ u_2^{(k)} \end{pmatrix}$ be the kth iterate obtained from the Gauss-Seidel iteration. We have

$$u_1^{(k+1)} - u_1^* = A(\alpha)S(\alpha)[u_1^{(k)} - u_1^*] \text{ and } u_2^{(k+1)} - u_2^* = S(\alpha)A(\alpha)[u_2^{(k)} - u_2^*].$$

Since $\|A(\alpha)S(\alpha)\| \leq \beta(\alpha)$ and $\|S(\alpha)A(\alpha)\| \leq \beta(\alpha)$,

$$\|u_1^{(k+1)} - u_1^*\| \leq \beta(\alpha)\|u_1^{(k)} - u_1^*\| \text{ and } \|u_2^{(k+1)} - u_2^*\| \leq \beta(\alpha)\|u_2^{(k)} - u_2^*\|.$$

Therefore, the solution $\mathbf{u}^*$ verifies

$$\||\mathbf{u}^{(k+1)} - \mathbf{u}^*\|| = \||L_1(\alpha)(\mathbf{u}^{(k)} - \mathbf{u}^*)\|| \leq \beta(\alpha)\||\mathbf{u}^{(k)} - \mathbf{u}^*\|| < \||\mathbf{u}^{(k)} - \mathbf{u}^*\||.$$

It follows that $\||L_1(\alpha)\|| \leq \beta(\alpha) < 1$ and the Gauss-Seidel iteration method converges. The Gauss-Seidel and SAS iterations methods have the same bound for their contraction factors. The equivalence between the two methods follows from the fact that the SAS method is the Gauss-Seidel method apply to system (103) and at each iteration k, the norm $\|.\|$ of the second component of the error vector function applying the Gauss-Seidel iteration to (108) is the same as the $\|.\|_{A(\alpha)}$ of the error function applying the SAS iteration to the initial problem (62). $\qquad\square$

Since the Gauss-Seidel method is convergent and is equivalent to the SAS iteration, there exists a value θ for which the number of iteration of the SOR method is less or equal to the number of the SAS iterations. Thus the SOR iteration method can accelerate the SAS iteration.

The results of Theorem 3.3.6 and Theorem 3.3.7 hold. Moreover, from relations (50) and (111), we deduce that the spectral radius of $\mathbf{J}(\alpha)$ verifies

$$r(\mathbf{J}(\alpha)) < \sqrt{\beta(\alpha)}. \tag{112}$$

Therefore, if $\sigma(\mathbf{J}(\alpha)) \subset \mathbb{R}$, the optimal SOR parameter can be approximated by

$$\theta_a = \frac{2}{1 + \sqrt{1 - \beta(\alpha)}}. \tag{113}$$

3.4.6 Discretization and Numerical Results.

Here, we investigate the numerical simulations for one dimensional slab geometry of width $b - a$, two dimensional case and one dimensional spherical geometry.

One Dimensional Case

For the discretization of the problem (102), we introduce a finite set of J discrete angular directions $\{\mu_k\}_{l=1}^J \in [-1, 1]$, which are nonzero and symmetric about the origin. Using the discrete directions μ_l ($l = 1, 2, \ldots, J$), the semi-discrete formulation of the problem (102) can be stated as follows:

$$\begin{cases} \frac{\mu_l \partial f_l}{\partial x} + \alpha f_l = \varphi_k, \text{ in } (a, b) \\ f(a, \mu_l) = f(b, -\mu_l) = 0, \mu_l > 0 \end{cases}, \tag{114}$$

where $f_l(x) = f(x, \mu_l)$, $l = 1, 2, \ldots, J$.

The fully discrete form of the problem (102) is obtained by discretizing the equation (114). The numerical grid of the spatial domain is defined by:

$$D_h = \{x_i, 0 \le i \le N\}, \tag{115}$$

where $x_0 = a$, $x_i = x_{i-1} + (\Delta x)_i$ ($0 < i < N$), $x_N = b$ and $h = \max\limits_{0 < i \le N} (\Delta x)_i$. The cell center grid points are $x_{i+\frac{1}{2}} = \frac{x_{i+1} + x_i}{2}$ and $(\Delta x)_{i+\frac{1}{2}} = x_{i+1} - x_i = h_i$.

Using the difference method based on control volume approach and cell averaging, a fully discrete form of the transport problem (102) can be written as follows:

$$\begin{cases} a_{l,i} f_{l,i+1} + b_{l,i} f_{f,i} = \varphi_{l,i+\frac{1}{2}}; 0 \le i \le N - 1 \\ f_{l,0} = 0 \end{cases} \quad \text{for } \mu_l > 0; \tag{116}$$

$$\begin{cases} b_{l,i} f_{l,i+1} + a_{l,i} f_{l,i} = \varphi_{l,i+\frac{1}{2}}; 0 \le i \le N - 1 \\ f_{l,N} = 0 \end{cases} \quad \text{for } \mu_l < 0, \tag{117}$$

where $a_{l,i} = \frac{|\mu_l|}{h_i} + \frac{\alpha}{2}$ and $b_{l,i} = -\frac{|\mu_l|}{h_i} + \frac{\alpha}{2}$. The matrices of the discrete systems (116) and (117) are sparse. These systems are solved exactly using forward substitution for $\mu_l > 0$ and backward substitution for $\mu < 0$, since their matrices are of triangular form. We present numerical results from the application of the SAS method on three example problems.

Two Dimensional Case

For the solution of (102) we introduce a finite set of J discrete angular directions $\Omega_J = \{\omega_i = (\mu_i, \eta_i, \xi_i)\}_{i=1}^{i=J} \subset S^1$, which are nonzero and symmetric about the origin.

Using the discrete directions of the set Ω_J, the semi-discrete formulation of the problem (102) reads:

$$\begin{cases} \mu_l \frac{\partial f_l}{\partial x} + \eta_l \frac{\partial f_l}{\partial y} + \alpha f_l = \varphi_l, & \text{in } D \times \Omega_J \\ f_l(\mathbf{x}) = 0, & \text{in } \partial D_- \times \Omega_J. \end{cases} \tag{118}$$

where for a given function f, $f_l(x, y) = f(x, y, \mu_l, \eta_l)$ $(1 \leq l \leq J)$.

We assume that D is the rectangular domain defined by $D =]a, b[\times]c, d[$. The numerical grid is defined by:

$$D_h = \{(x_i, y_j), 0 \leq i \leq N, 0 \leq i \leq M\}, \tag{119}$$

where $x_0 = a$, $x_i = x_{i-1} + (\Delta x)_i$, $x_N = b$, $y_0 = c$, $y_j = y_{j-1} + (\Delta y)_j$, $y_M = d$ and $h = \max_{ij} ((\Delta x)_i, (\Delta y)_j)$. The cell center grid points are defined as:

$$x_{i+\frac{1}{2}} = \frac{x_{i+1} - x_i}{2}, \quad y_{j+\frac{1}{2}} = \frac{y_{j+1} - y_j}{2}, \quad (\Delta x)_{i+\frac{1}{2}} = x_{i+1} - x_i \text{ and } (\Delta y)_{j+\frac{1}{2}} = y_{j+1} - y_j.$$

Denoting by f_{ij} the approximation of the function f at the node (x_i, y_j), the value of f at the cell center is approximated by:

$$f_{i+\frac{1}{2}j+\frac{1}{2}} = \frac{f_{ij} + f_{i+1j} + f_{ij+1} + f_{i+1j+1}}{4}. \tag{120}$$

Using the centered difference method for the approximation of derivative, a fully discrete form of problem (102) reads :

$$\mu_l \frac{f_{l,i+1j+\frac{1}{2}} - f_{l,ij+\frac{1}{2}}}{(\Delta x)_{i+\frac{1}{2}}} + \eta_l \frac{f_{l,i+\frac{1}{2}j+1} - f_{l,i+\frac{1}{2}j}}{(\Delta y)_{i+\frac{1}{2}}} + \alpha f_{l,i+\frac{1}{2}j+\frac{1}{2}} = \varphi_{l,i+\frac{1}{2}j+\frac{1}{2}}. \tag{121}$$

Using the relations $f_{ij+\frac{1}{2}} = \frac{f_{ij}+f_{ij+1}}{2}$ and $f_{i+\frac{1}{2}j} = \frac{f_{ij}+f_{i+1j}}{2}$, system (121) writes:

$$a_{l,ij} f_{l,ij} + b_{l,ij} f_{l,ij+1} + c_{l,ij} f_{l,i+1j} + d_{l,ij} f_{l,i+1j+1} = \varphi_{l,i+\frac{1}{2}j+\frac{1}{2}}, \tag{122}$$

where $a_{l,ij} = \left(-\theta_{l,i} - \beta_{l,j} + \frac{\alpha}{4}\right)$, $b_{l,ij} = \left(-\theta_{l,i} + \beta_{l,j} + \frac{\alpha}{4}\right)$, $c_{l,ij} = \left(\theta_{l,i} - \beta_{l,j} + \frac{\alpha}{4}\right)$ and $d_{l,ij} = \left(\theta_{l,i} + \beta_{l,j} + \frac{\alpha}{4}\right)$; with $\theta_{l,i} = \frac{\mu_l}{2(\Delta x)_{i+\frac{1}{2}}}$ and $\beta_{l,j} = \frac{\eta_l}{2(\Delta y)_{j+\frac{1}{2}}}$.

The system (122) is explicitly solved using the sweeping method.

Stability result: Let α be a positive constant. We have $\|(A + \alpha I)^{-1}\| \leq \frac{1}{\alpha}$ and the solution f of equation (102) verifies

$$\|f\| \leq \frac{1}{\alpha} \|\varphi\|. \tag{123}$$

It follows using piecewise bilinear approximation in space and piecewise constant approximation in angle, that the solution $\{f_{l,i+1/2j+1/2}\}$ of (121) satisfies

$$|f_h|_h \leq \frac{1}{\alpha} |\varphi_h|_h, \qquad (124)$$

where $|.|_h$ is the following discrete analogue of the norm $\|.\|$:

$$|f_h|_h^2 = \sum_{i=0}^{N-1} \sum_{j=0}^{M-1} \sum_{l=1}^{J} (\Delta x)_{i+\frac{1}{2}} (\Delta y)_{i+\frac{1}{2}} W_l f_{l,i+1/2j+1/2}^2,$$

with W_l being the weight associated to the angular node $(\mu_l, \eta_l) \in \Omega_J$. Thus the discrete scheme (121) is stable. Moreover this scheme is of second order of accuracy and is generally not monotonic. The monotonicity of the scheme can be achieved in the limit case where $h \to 0$ (see [34]).

Numerical Results

We took particular data for which an exact solution of problem (4)-(4) is known in each case. For the iterative methods tested here, the iterations are stopped when the relative error $\frac{\|U - U_{exact}\|_2}{\|U_{exact}\|_2}$ is less than a prescribed $\epsilon > 0$.

Let N_s and N_a denote the total number of spatial grid points an discrete angular directions respectively. At each iteration of the SAS method, we have the following computational cost in one and two dimensional space isotropic cases: The cost of the sweeping algorithm is $O(N_s N_a)$ floating point operations (specifically $4 N_s N_a$ for 1-D and $8 N_s N_a$ for 2-D). The solution of equation (100) using (95) needs $O(N_s N_a)$ floating point operations. Updating the right hand side took $O(N_s N_a)$ floating point operations. The overall computational cost at each iteration is $O(N_s N_a)$. In the anisotropic case, the computational cost at each iteration is $O(n_{CG} N_s N_a^2)$, where n_{CG} is the number of iterations necessary for the convergence of CG algorithm. It decreases with the number of SAS iteration. In both case, there is no matrix storage.

Slab geometry case: $a = 0, b = 1$; for $x \in (0,1)$ and $\mu, \mu' \in [-1,1]$, we set $\sigma(x) = \sigma, \kappa(x, \mu, \mu') = \sigma c/2$ and

$$q(x, \mu) = \begin{cases} \mu^2 + \sigma \mu x - \frac{\sigma c}{4} & \mu > 0 \\ \mu^2 + \sigma \mu (x - 1) - \frac{\sigma c}{4} & \mu < 0 \end{cases}.$$

The exact solution of this problem is given by:

$$\psi(x, \mu) = \begin{cases} \mu x & \mu > 0 \\ \mu(x - 1) & \mu < 0 \end{cases}.$$

84

For the numerical test, we take $N = 10$ and $J = 10$. We study the behavior of the SAS method with respect to parameters α, c and σ. The theoretical value of the SAS iteration optimal parameter is $\alpha_t = \sigma(1 - c)$. The SAS iterations was applied to the example problem for several values of c and σ. For fixed c and σ, the numerical optimal value of α) can be localized in the interval $[\sigma(1 - c), \sigma(1 - c/2)]$. We take

$$
\alpha = \begin{cases}
\sigma(1 - 19c/32), & 0 < c < 0.9 \\
\sigma(1 - 23c/32), & 0.9 \le c < 0.97 \\
1, & 0.97 \le c \le 1
\end{cases}
$$

We compare the number of iteration and the CPU time of the standard Source Iteration (SI) method and the SAS method. There is two sets of tests: one for fixed c, another for fixed σ. As shown by Figure 12 to Figure 14 , the SAS method converges faster than the standard SI method, particularly for large values σ and c. For $c = 0.99$, we compare the σ dependence of the iterative methods used here, we can see on Figure 24 that the number of iteration of the SAS method is the same as that of the ISAS method using the Conjugate Gradient (ISAS(CG)) method for the first inner equation. It can also be seen that the computing times of SAS and ISAS(CG) are roughly the same and are less than that of SI method. This observation remains true for very large values of σ (Figure 14). We observe that the SAS method is faster than SI method, and converges for $c = 1$.

Two dimensional case: $D =]0, 1[\times]0, 1[$; for $x = (x_1, x_2) \in D$ and $\Omega = (\mu, \eta) \in$ $B(0, 1) = \{\Omega \in \mathbb{R}^2, \|\Omega\|_2 < 1\}$, we set $\sigma(x) = \sigma$, $\kappa(x, \Omega, \Omega') = \frac{\sigma c}{\pi}$ and

$$
q(x, \mu) = \begin{cases}
\mu x_2 + \eta x_1 + \sigma x_1 x_2 - \frac{\sigma c}{4} & \mu > 0\ \eta > 0 \\
-\mu x_2 + \eta(1 - x_1) + \sigma(1 - x_1)x_2 - \frac{\sigma c}{4} & \mu < 0\ \eta > 0 \\
-\mu(1 - x_2) - \eta(1 - x_1) + \sigma(1 - x_1)(1 - x_2) - \frac{\sigma c}{4} & \mu < 0\ \eta < 0 \\
\mu(1 - x_2) - \eta x_1 + \sigma x_1(1 - x_2) - \frac{\sigma c}{4} & \mu > 0\ \eta < 0
\end{cases}.
$$

The exact solution of this test problem is given by:

$$
\psi(x, \mu) = \begin{cases}
x_1 x_2 & \mu > 0\ \eta > 0 \\
(1 - x_1)x_2 & \mu < 0\ \eta > 0 \\
(1 - x_1)(1 - x_2) & \mu < 0\ \eta < 0 \\
x_1(1 - x_2) & \mu > 0\ \eta < 0
\end{cases}.
$$

For the numerical tests, we take $\Delta x = \Delta y = \frac{1}{10}$ and $J = 100$. We perform the the same set of test as in the case of slab geometry, using the previous values of α for SAS iteration. For $c = 0.98$, we compare the σ dependence of the methods tested here. It can be seen (Figure 15) that SAS method is efficient compare to SI method, even for large values of σ

85

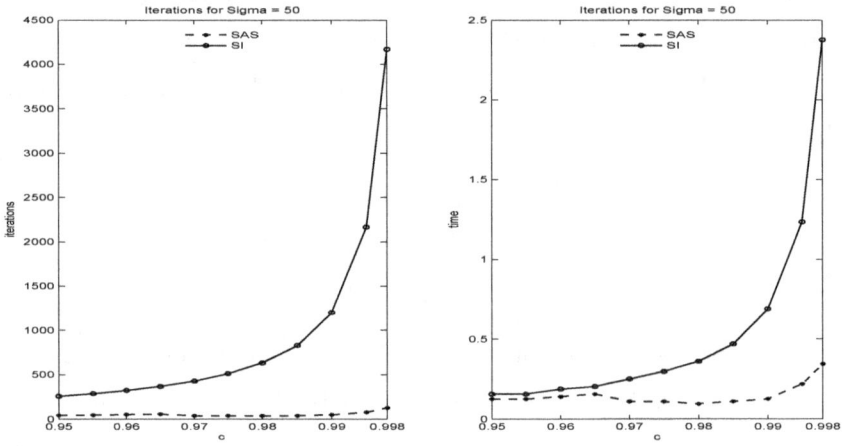

Figure 12: Comparison at fixed $\sigma = 50$ of Source Iteration and SAS in slab geometry for values of c close to 1 ($\epsilon = 1E - 06$): (left) Number of iterations; (right) CPU time.

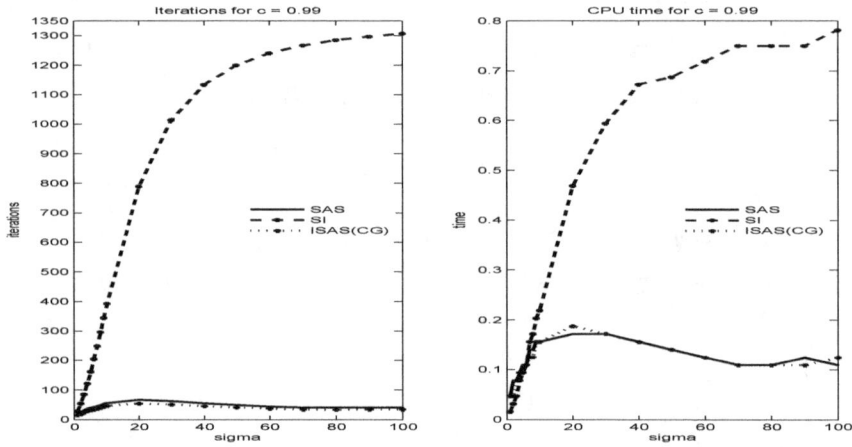

Figure 13: Comparison at fixed c=0.99 of Source Iteration, SAS and ISAS(CG) in slab geometry ($\epsilon = 1E - 06$): (left) Number of iterations; (right) CPU time.

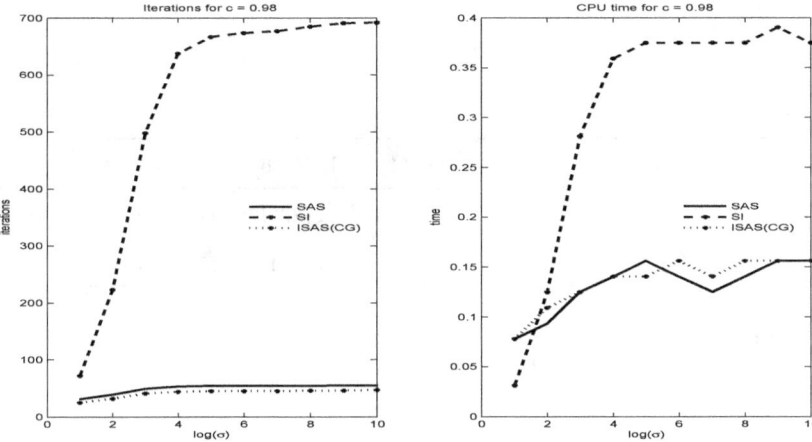

Figure 14: Comparison at fixed $c = 0.98$ of Source Iteration, SAS and ISAS(CG) in slab geometry for large values of σ ($\epsilon = 1E - 06$): (left) Number of iterations; (right) CPU time.

(Figure 25). We also compare the c dependence of the two iteration methods for $\sigma = 50$, with c near one (Figure 17). We can see that the SAS algorithm is still more efficient. We plot in Figure 18 the convergence rate of the two methods at $c = 0.5$ and $c = 1$ for several values of σ. We can see that SAS algorithm converge for $c = 1$, even for very large values of σ. As in the case of slab geometry, the SAS algorithm is more efficient than the Source Iteration algorithm, particularly for the critical cases (c close to 1 and/or large σ).

Additionally, we present in Table 1 comparative numerical results (number of iteration and cpu time) of SAS and IDR(s) (s denoting the dimension of the Krylov subspace), SSOR preconditioned GMRES (fgmres), BiCGStab iterative algorithm with Gauss-Seidel preconditioner (Bicgstab) and a spatial multigrid method (MG). For this set of tests, We take $\Delta x = \Delta y = \frac{1}{8}$ and $J = 100$. The matrix G has 2562500 non zeros entries. The iterations are stopped when the relative error $\frac{\|B-GU\|_2}{\|B\|_2}$ is less than $1E - 05$, where B denotes the right hand side of the discrete system. The convergence behavior (relative residual as function of iteration) of the SAS, MG and fgmres are plotted in Figure 19 and Figure 20 for the cases $c = 0.5$ and $c = 0.98$ respectively. It can be observed that the SAS method is efficient in both cases as compare to other methods tested here.

87

	MG	SAS	IDR(4)	IDR(16)	Bicgstab	fgmres
$c = 0.5$	23(107.87)	7(0.95)	87(5.28)	82(2.29)	41(11.28)	32(5.01)
$c = 0.98$	> 200	33(4.32)	125(7.56)	106(7.15)	77(20.97)	95(15.34)

Table 1: Iterations number and CPU time in s (in bracket) at $\sigma = 100$

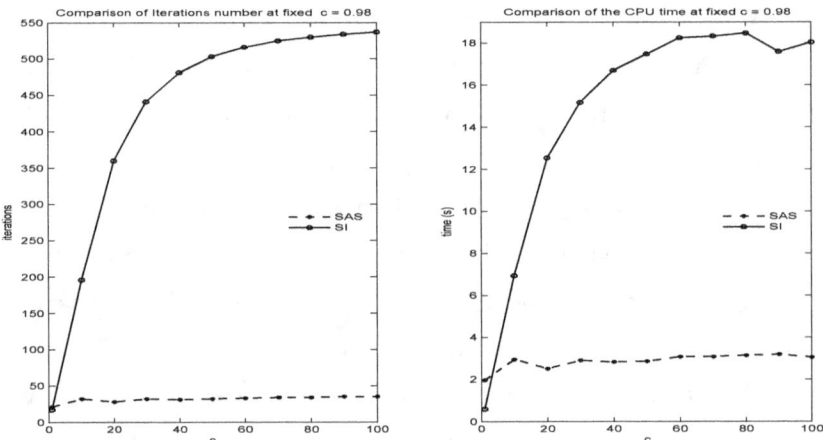

Figure 15: Comparison of the SI and SAS methods in 2D case at fixed $c = 0.98$ ($\epsilon = 1E - 5$): (left) number of iterations; (right) CPU time.

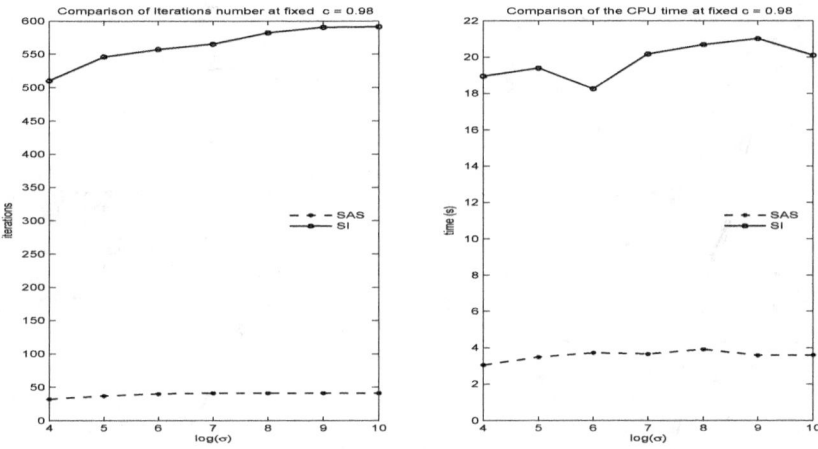

Figure 16: Comparison of the SI and SAS methods in 2D case at fixed $c = 0.98$, for large σ ($\epsilon = 1E - 5$): (left) number of iterations; (right) CPU time.

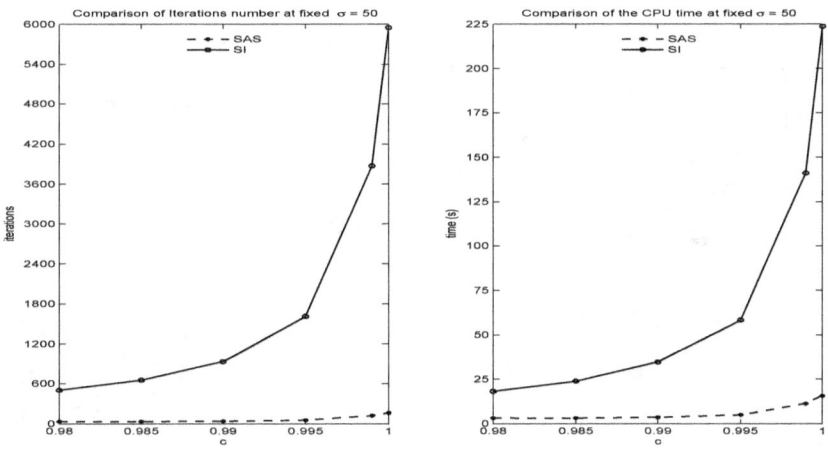

Figure 17: Comparison of the SI and SAS methods in 2D case at fixed $\sigma = 50$, for $c \approx 1$ ($\epsilon = 1E - 05$): (left) number of iterations; (right) CPU time.

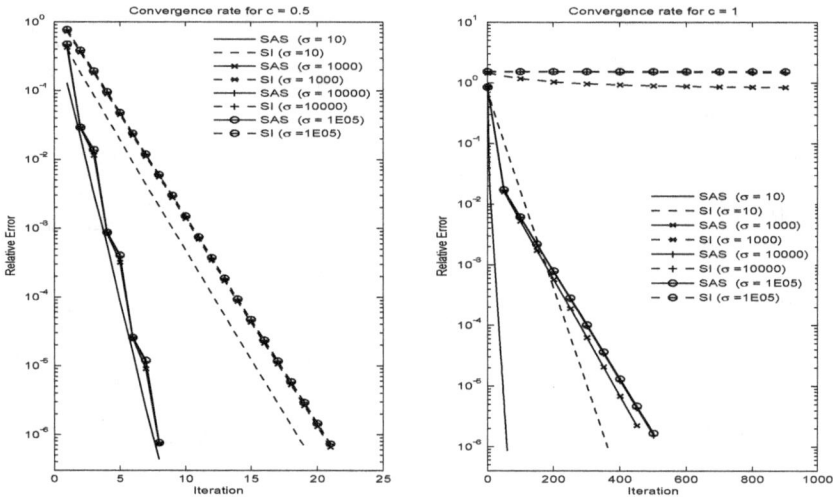

Figure 18: Comparison of the convergence rate in 2D case at fixed c of Source Iteration and SAS iteration for several values of σ: (left) $c = 0.5$ and (right) $c = 1$.

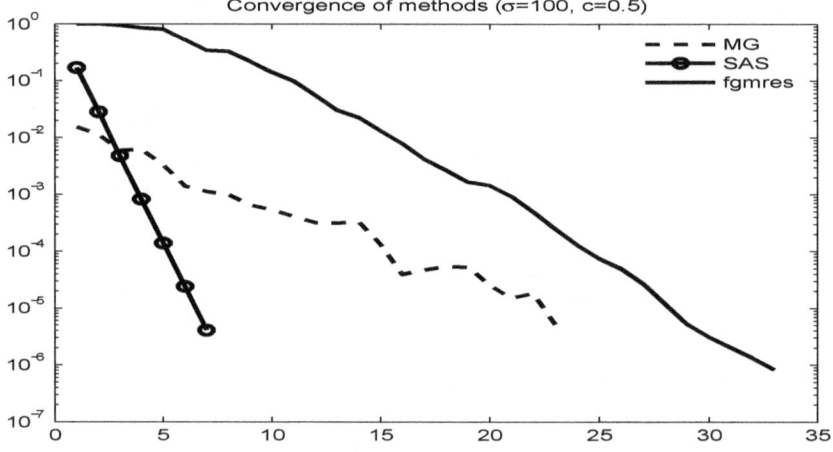

Figure 19: Convergence of the SAS, fgmres and MG ($c = 0.5$) methods at fixed $\sigma = 100$.

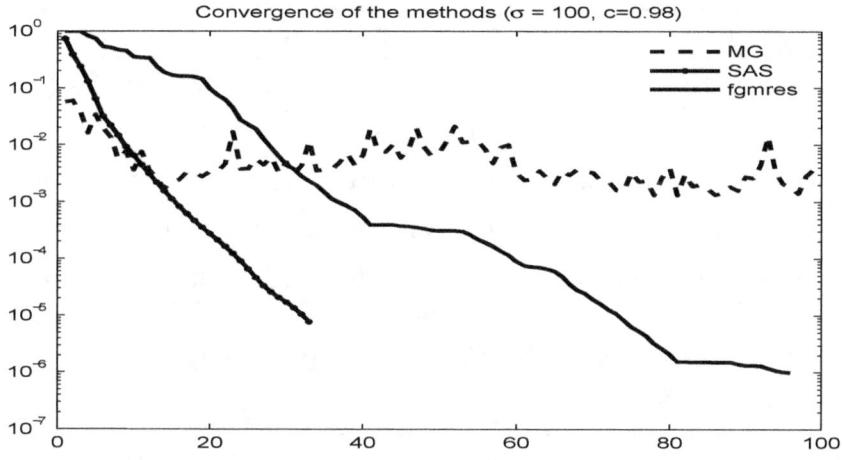

Figure 20: Convergence of the SAS, fgmres and MG ($c = 0.98$) methods at fixed $\sigma = 100$.

Numerical results of the SOR acceleration of the SAS iteration in Two dimensional case

The spatial and angular discretization are carried out by the DSN scheme. The spatial mesh size is $h_x = h_y = 1/10$ and we divided the unit disc in 100 regions. We Study the behavior of Source Iteration (SI), Jacobi, SAS Gauss-Seidel (GS) and SOR (we set $\omega = 0.93$) with respect to σ and c. At fixed $\sigma = 100$, we set $\alpha = \sigma(1-c)$ and compare the c-dependence of the iterative methods. As shown by Table 2, all the method converge. It is observed that SOR method is efficient compare to the other methods. At fixed $c = 0.95$, we set $\alpha = 10$ and compare the σ-dependence of the method. The numerical results in Table 3 shows the efficiency of the SOR method. It can be observed that SAS is comparable to GS, which is two time faster than Jacobi method. This confirms the theoretical results. The Figure 21 plots the convergence behavior of SAS and SOR methods for $\alpha = 5$ at fixed $\sigma = 100$ and $c = 0.99$, with $\omega = 0.93$. We set $\alpha_1 = \sigma$, $\alpha_2 = \sigma\sqrt{(1-c)}$, $\omega_a = \frac{2}{1+\sqrt{1-\beta(\alpha_1)}}$ for $\sigma > 15$ and $\omega_a = 0.93$ for $\sigma \geq 15$. Comparative numerical results of SAS ($\alpha = \alpha_1$), SAS ($\alpha = \alpha_2$), SOR ($\alpha = \alpha_1, \omega = \omega_a$) and SOR ($\alpha = \alpha_2, \omega = 0.93$) methods are plotted in Figure 24 at fixed $c = 0.99$ ($\sigma \in [1, 100]$) and in Figure 25 at fixed $\sigma = 100$ ($c \in [0.10.99]$). It is observed that the SOR ($\alpha = \alpha_2, \omega = 0.93$) method is efficient compare to the other methods.

91

c	0.1	0.3	0.5	0.7	0.8	0.9	0.95
SI	7(0.23)	13(0.42)	22(0.73)	42(1.39)	66(2.20)	138(4.92)	281(9.64)
SAS	5(0.50)	8(0.71)	12(1.09)	21(1.87)	31(2.90)	53(4.84)	56(4.95)
Jacobi	8(0.67)	14(1.15)	22(1.92)	40(3.34)	60(4.87)	104(9.07)	110(9.20)
GS	4(0.32)	7(0.59)	11(0.96)	20(1.70)	30(2.48)	52(4.42)	55(4.67)
SOR	5(0.39)	5(0.42)	6(0.54)	11(0.92)	14(1.17)	18(1.48)	24(2.15)

Table 2: Iterations number and CPU time in s (in bracket) at fixed $\sigma = 100$.

σ	20	40	60	70	80	90	100
SI	235(7.35)	268(8.43)	276(8.84)	278(8.71)	279(8.79)	280(8.81)	281(8.78)
SAS	62(5.04)	36(2.96)	35(2.87)	40(3.26)	44(3.70)	49(3.93)	53(4.34)
Jacobi	121(9.39)	67(5.20)	68(5.25)	78(6.00)	86(6.65)	96(7.53)	104(8.00)
GS	61(4.82)	35(2.73)	34(2.60)	39(3.09)	43(3.34)	48(3.67)	52(4.09)
SOR	71(5.46)	41(3.15)	28(2.14)	24(1.89)	21(1.60)	19(1.48)	19(1.50)

Table 3: Iterations number and CPU time in s (in bracket) at fixed $c = 0.95$ ($\alpha = 10$).

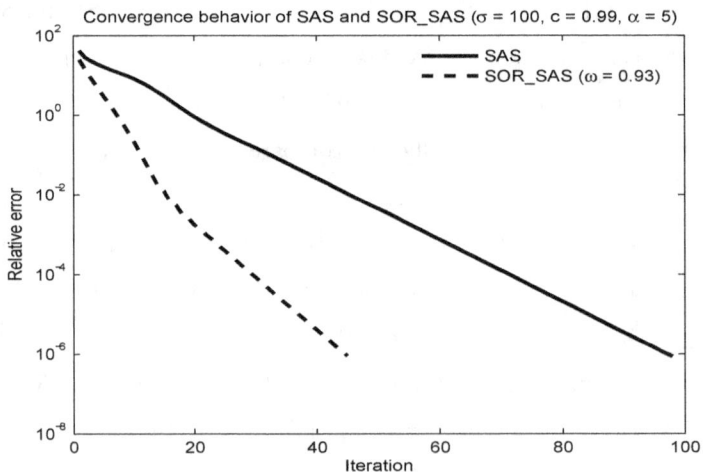

Figure 21: Convergence behavior of SAS and SOR ($\omega = 0.93$) methods at fixed $\sigma = 100$, $c = 0.99$ and $\alpha = 5$.

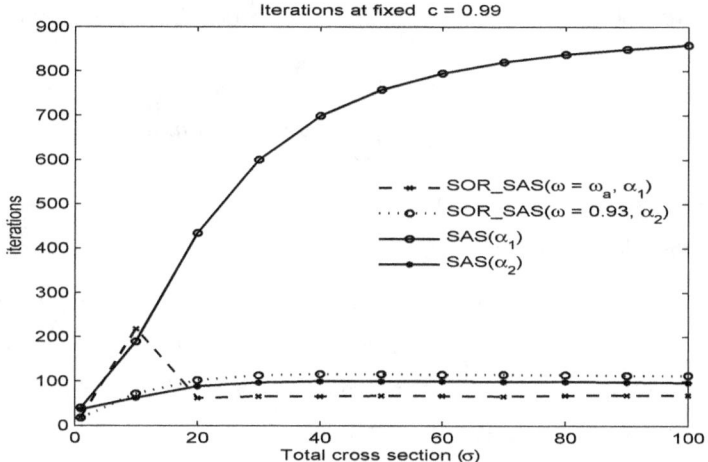

Figure 22: Comparison of the SAS ($\alpha = \alpha_1$), SAS ($\alpha = \alpha_2$), SOR ($\alpha = \alpha_1, \omega = \omega_a$) and SOR ($\alpha = \alpha_2, \omega = 0.93$) methods at fixed $c = 0.99$, for $\sigma \in [1, 100]$($\epsilon = 1E - 08$).

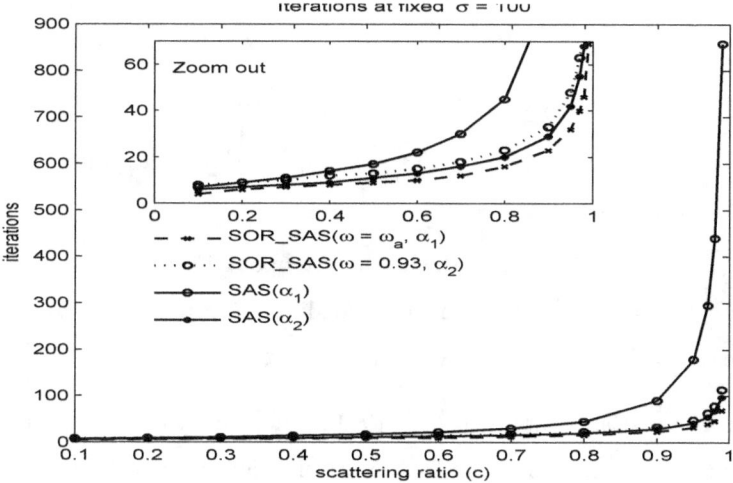

Figure 23: Comparison of the SAS ($\alpha = \alpha_1$), SAS ($\alpha = \alpha_2$), SOR ($\alpha = \alpha_1, \omega = \omega_a$) and SOR ($\alpha = \alpha_2, \omega = 0.93$) methods at fixed $\sigma = 100$, for $c \in [0.1, 0.99]$ ($\epsilon = 1E - 08$).

The spherical geometry case

Discretization. Let $\Omega = (0, R) \times (-1, 1)$. We consider the following triangulation of Ω:

$$\overline{\Omega} = \bigcup_{i,j}([r_i, r_{i+1}] \times [\mu_j, \mu_{j+1}]) = \bigcup_{i,j} \Omega_{i,j} \quad 0 \le i \le N - 1 \text{ and } -J \le j \le J - 1,$$

where $0 = r_0 < r_1 < \ldots < r_N = R$ and $-1 = \mu_{-J} < \mu_{-J+1} < \ldots < \mu_J = 1$ with $\mu_{-j} = -\mu_j$. We also consider the following nodes: $r_{i+\frac{1}{2}} = \theta_{i+\frac{1}{2}} r_{i+1} + (1 - \theta_{i+\frac{1}{2}}) r_i$ and $\mu_{j+\frac{1}{2}} = \gamma_{j+\frac{1}{2}} \mu_{j+1} + (1 - \gamma_{j+\frac{1}{2}}) \mu_i$ where $\theta_{i+\frac{1}{2}} \in (0, 1)$ $(0 \le i \le N - 1)$; $\gamma_{j+\frac{1}{2}} \in (0, 1)$ and $\gamma_{j+\frac{1}{2}} = 1 - \gamma_{-(j+\frac{1}{2})}$ $(-J \le j \le J - 1)$. Let $U_{x,y}$ denotes the approximate value of the flux at (r_x, μ_y). We suppose that in Ω_{ij}, the approximate flux is a polynomial of degree ≤ 1 in r and μ. Therefore, we have the following relations:

$$\begin{cases} \theta_{i+\frac{1}{2}} U_{i+1,j+\frac{1}{2}} + (1 - \theta_{i+\frac{1}{2}}) U_{i,j+\frac{1}{2}} = U_{i+\frac{1}{2},j+\frac{1}{2}} \\ \gamma_{j+\frac{1}{2}} U_{i+\frac{1}{2},j} + (1 - \gamma_{j+\frac{1}{2}}) U_{i+\frac{1}{2},j} = U_{i+\frac{1}{2},j+\frac{1}{2}} \end{cases} . \tag{125}$$

Using the DSN difference scheme [34], the discrete form of the equation (102) reads:

$$\begin{cases} \eta_{j+\frac{1}{2}}(r_{i+1}^2 U_{i+1,j+\frac{1}{2}} - r_i^2 U_{i,j+\frac{1}{2}}) + \lambda_{i+\frac{1}{2}}(\beta_{j+1} U_{i+\frac{1}{2},j+1} - \beta_j U_{i+\frac{1}{2},j}) + \\ \alpha \nu_{i+\frac{1}{2},j+\frac{1}{2}} U_{i+\frac{1}{2},j+\frac{1}{2}} = \nu_{i+\frac{1}{2},j+\frac{1}{2}} f_{i+\frac{1}{2},j+\frac{1}{2}} \\ 0 \le i \le N - 1; -J \le j \le J - 1 \end{cases} , \tag{126}$$

where $\eta_{j+\frac{1}{2}} = \omega_{j+\frac{1}{2}} \mu_{j+\frac{1}{2}}$, $\lambda_{i+\frac{1}{2}} = w_{i+\frac{1}{2}} r_{i+\frac{1}{2}}$, $\nu_{i+\frac{1}{2},j+\frac{1}{2}} = \omega_{j+\frac{1}{2}} w_{i+\frac{1}{2}} r_{i+\frac{1}{2}}^2$ and

$$\begin{cases} r_{i+1}^2 - r_i^2 = 2 w_{i+\frac{1}{2}} r_{i+\frac{1}{2}}, \quad r_{i+1}^3 - r_i^3 = 3 w_{i+\frac{1}{2}} r_{i+\frac{1}{2}}^2 \\ \beta_{j+1} \mu_{j+1} - \beta_j \mu_j = \omega_{j+\frac{1}{2}}(1 - 3\mu_{j+\frac{1}{2}}^2), \quad \beta_{j+1} - \beta_j = -2\mu_{j+\frac{1}{2}} \omega_{j+\frac{1}{2}} \\ \omega_{j+\frac{1}{2}} = \mu_{j+1} - \mu_j, \quad \beta_J = 0 \end{cases} , \tag{127}$$

The system (126) is completed with the linear system obtained from discretization of the equation (102) at $\mu = -1$ and the symmetry condition at $r = 0$. The resulting system is solved explicitly to obtain $U_{i,j+\frac{1}{2}}$, $U_{i+\frac{1}{2},j}$ and $U_{i+\frac{1}{2},j+\frac{1}{2}}$.

Numerical results. We present numerical results from the application of the ISAS method on an example problem. We took particular data for which an exact solution u is known: $\sigma(r) = \sigma$, $\kappa(r, \mu, \mu') = \frac{\sigma c}{2}$, $(0 < c < 1)$, $R = 1$

$$q(r, \mu) = \sigma(1 - c)(1 - r) - \mu; \quad u(r, \mu) = 1 - r$$

and we compared the speed (number of iteration and CPU time) of SAS, SOR acceleration and standard SI algorithms at fixed c and at fixed σ. For iterative methods tested here,

94

the iterations are stopped when the relative error $\frac{\|U-U^{(k)}\|_2}{\|U\|_2}$ is less than a prescribed ϵ. For this problem, the theoretical value of the SAS iteration optimal parameter is $\alpha_t = \sigma(1-c)$. For the SAS iteration, we set $\alpha = \sigma$. The spatial mesh size is $h = 1/10$ and the angular mesh size is $\tau = 1/10$.

The Figure 24 represents as function of σ ($10 \le \sigma \le 100$) the number of iteration and the CPU time at $c = 0.98$ of SAS, SI and SOR ($\omega = 1.2$) algorithms for $\epsilon = 1E - 05$. The Figure 25 represents as function of c ($0 < c < 1$) the number of iteration and the CPU time at $\sigma = 100$ of SAS, SI and SOR ($\omega = 1.3$) algorithms for $\epsilon = 1E - 05$. The Figure 26 represents as function of σ (for large σ) the number of iteration and the CPU time at $c = 0.98$ corresponding to SAS, SI and SOR ($\omega = 1.6$) algorithm for $\epsilon = 1E - 05$. The Figure 27 represents as function of c ($c \approx 1$) the number of iteration and the CPU time at $\sigma = 100$ corresponding to SAS, SI and SOR ($\omega = 1.7$) for $\epsilon = 5E - 04$. The Figure 28 represents the convergence rate as function of the number of iteration at $c = 0.5$ ($\sigma \in \{10, 100, 500\}$) and at $c = 1$ ($\sigma \in \{10, 50, 100\}$). It appears from these tests that the SAS algorithm is efficient compare to SI algorithm and can be accelerated by the SOR scheme.

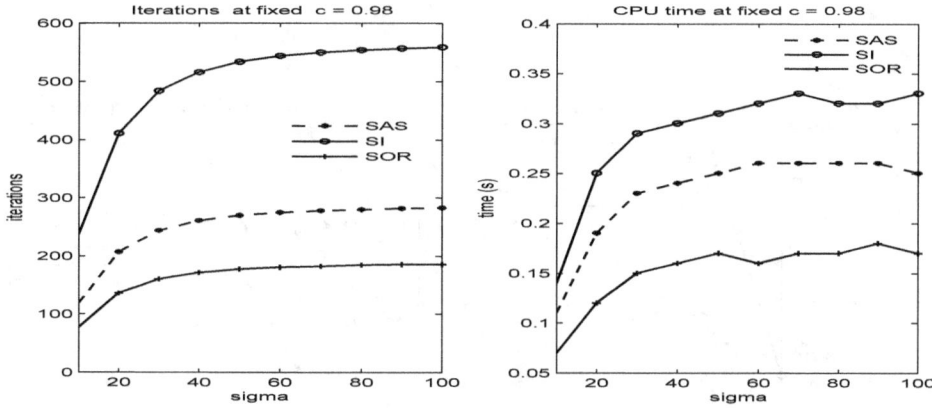

Figure 24: Comparison of the SI, SAS and SOR methods at fixed $c = 0.98$ ($\epsilon = 10^{-5}$): (left) number of iterations; (right) CPU time.

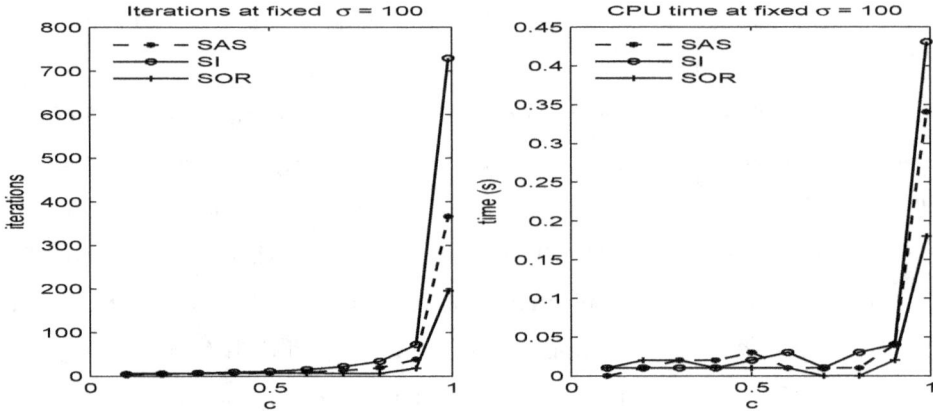

Figure 25: Comparison of the SI, SAS and SOR methods at fixed $\sigma = 100$ ($\epsilon = 10^{-5}$): (left) number of iterations; (right) CPU time.

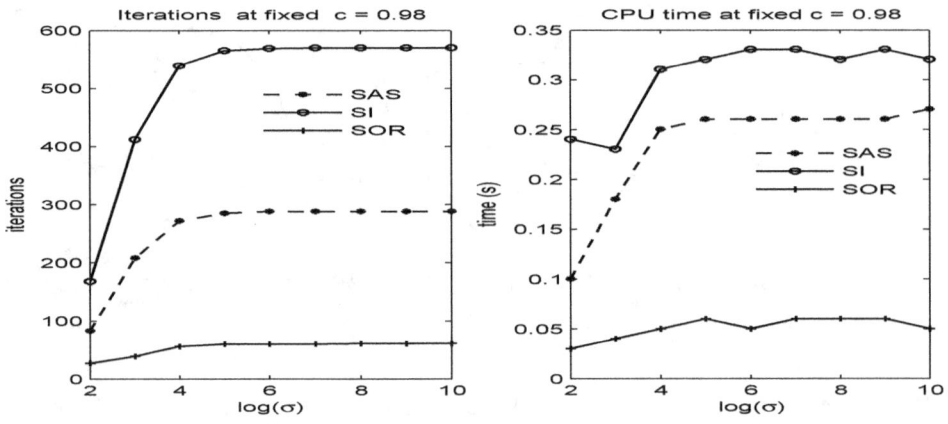

Figure 26: Comparison of the SI, SAS and SOR methods at fixed $c = 0.98$, for large σ ($\epsilon = 10^{-5}$): (left) number of iterations; (right) CPU time.

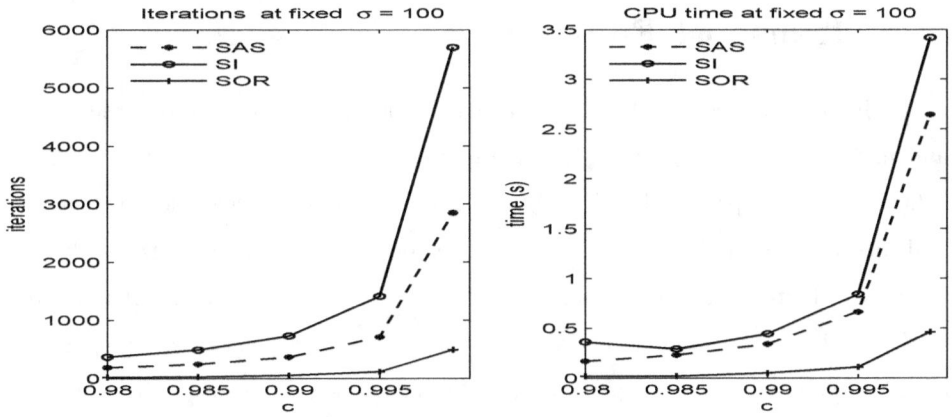

Figure 27: Comparison of the SI, SAS and SOR methods at fixed $\sigma = 50$, for $c \approx 1$ ($\epsilon = 5E - 04$): (left) number of iterations; (right) CPU time.

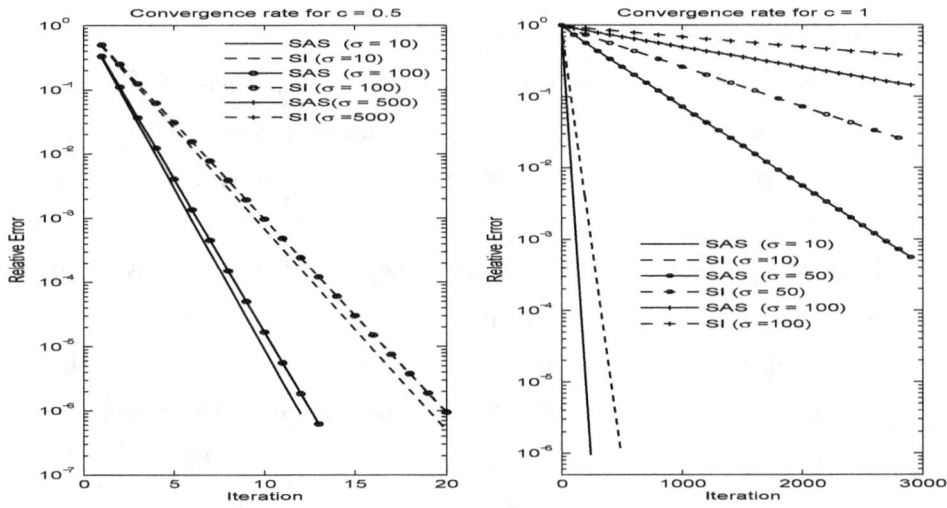

Figure 28: Comparison of the convergence rate at fixed c of the SI and SAS for several values of σ : (left) $c = 0.5$; (right) $c = 1$.

3.5 A Minimal Residual Iterative Solver for Neutron Transport Equation

In the previous section, ADI-like iterative method (see [59]) based on positive definite and m-accretive splitting for linear operator equations with operators admitting such splitting has been proposed and analyzed. This method converges unconditionally. The SOR acceleration of this method yields convergence results similar to those obtained in presence of finite dimensional systems with matrices possessing property A (see [50, 89]). In the particular case where the positive definite part of the linear equation operator is in addition self-adjoint, an upper bound for the contraction factor of the iterative method which depends solely on the spectrum of the self-adjoint part is derived. This method has been successfully applied to the neutron transport equation in slab and 2-D cartesian geometry and in 1-D spherical geometry.

The self-adjoint and m-accretive splitting leads to a fixed point problem where the operator is a 2 by 2 matrix of operators. An infinite dimensional adaptation of a minimal residual algorithm is then applied to solve the matrix operator equation. Theoretical analysis shows that the method converges unconditionally and an upper bound of the rate of residual decreasing which depends solely on the spectrum of the self-adjoint part of the operator is derived. The convergence of this solver and its Gauss-Seidel preconditioned version is numerically illustrated on a sample neutron transport problem in 2-D geometry where various test cases, including pure scattering and optically thick domains are considered.

The outline of this section is as follows. The description and the convergence properties of the SAS iteration method are given in subsection 3.5.1. In subsection 3.5.2, we present the minimal residual method and the convergence analysis. Subsection 3.5.3 deals with the application of the method to 2-D neutron transport equation and the numerical illustration.

3.5.1 The SAS Iteration Method

Let us consider a Hilbert space H with inner product $(.,.)$ and norm $\|.\|$ and let T be a linear operator on H with domain $\mathcal{D}(T)$ and range $\mathcal{R}(T) = H$. We denote by I, the identity operator. Suppose that we need to solve in $\mathcal{D}(T)$, the following problem

$$Tu = q, \tag{128}$$

where $q \in H$ is given and $u \in \mathcal{D}(T)$ is the unknown.

We assume that the operator T admits the following splitting:

$$T = S + A, \tag{129}$$

where S is a bounded self-adjoint positive definite operator and A is a m-Accretive operator. Therefore, the operator T is positive definite and equation (128) admits a unique solution in H.

We consider in $\mathcal{D}(T)$ the norm

$$\|u\|_{\mathcal{D}(T)} = \left(\|u\|^2 + \|Au\|^2 \right)^{\frac{1}{2}}. \tag{130}$$

Let α be a positive constant. The functional $\|.\|_{A(\alpha)}$ defined on $\mathcal{D}(T)$ by

$$\|u\|_{A(\alpha)} = \|(\alpha I + A)u\|, \tag{131}$$

is a norm on $\mathcal{D}(T)$ equivalent to $\|.\|_{\mathcal{D}(T)}$ (see [12]).

Let α be a positive constant. The following two-step splitting is obtained from (129):

$$\begin{cases} T = (\alpha I + S) - (\alpha I - A) \\ T = (\alpha I + A) - (\alpha I - S) \end{cases}. \tag{132}$$

The two-step splitting (132) leads to the following Self-adjoint and m-Accretive Splitting (SAS) iteration method (see [13]):

Given an initial guess $u^{(0)} \in D(T)$, for $k = 0, 1, \ldots$ until $\{u^{(k)}\}$ converges, calculate

$$\begin{cases} (\alpha I + S)u^{(k+\frac{1}{2})} = (\alpha I - A)u^{(k)} + q \\ (\alpha I + A)u^{(k+1)} = (\alpha I - S)u^{(k+\frac{1}{2})} + q \end{cases}. \tag{133}$$

From equation (133), we deduce that $u^{(k+1)}$ satisfies

$$(\alpha I + A)u^{(k+1)} = M(\alpha)(\alpha I + A)u^{(k)} + N(\alpha)q, \tag{134}$$

where

$$M(\alpha) = S_1(\alpha)A_1(\alpha) \text{ and } N(\alpha) = 2\alpha(\alpha I + S)^{-1}; \tag{135}$$

with

$$S_1(\alpha) = (\alpha I - S)(\alpha I + S)^{-1} \text{ and } A_1(\alpha) = (\alpha I - A)(\alpha I + A)^{-1}. \tag{136}$$

Therefore, the exact solution u^* of the problem (128) verifies

$$\|u^{(k+1)} - u^*\|_{A(\alpha)} \leq \|M(\alpha)\| \|u^{(k)} - u^*\|_{A(\alpha)}. \tag{137}$$

Let $\sigma(S)$ denotes the spectrum of S. Since A is m-accretive and S is self-adjoint, it is proved in [13, 15] that:

$$\|A_1(\alpha)\| \leq 1 \text{ and } \|S_1(\alpha)\| \leq \beta(\alpha) < 1, \tag{138}$$

where

$$\beta(\alpha) = \sup_{\lambda \in \sigma(S)} \left| \frac{\alpha - \lambda}{\alpha + \lambda} \right|. \tag{139}$$

Thus

$$\|M(\alpha)\| \leq \beta(\alpha) < 1. \tag{140}$$

It then follows that, the SAS iteration (133) converges unconditionally to the solution of (128) with respect to norm $\|.\|_\alpha$ and $\|.\|_{D(T)}$. Since for $u \in D(T)$, we have $\|u\| \leq \|u\|_{D(T)}$, the convergence of the SAS iteration with respect to the norm $\|.\|$ follows.

The theoretical optimal parameter α^* for the bound $\beta(\alpha)$ is $\alpha^* = \sqrt{\lambda_{min}\lambda_{max}}$ with λ_{min} and λ_{max} denoting respectively the lower and upper bound of $\sigma(S)$ (see [59]).

Each step of the SAS iterative method is constituted of two-half steps which require finding solutions of linear equations with operators $(\alpha I + S)$ and $(\alpha I + A)$. These linear operator equations can be solved approximately using appropriate methods with respect to the properties of each operators. This results in the inexact Self-adjoint and m-Accretive splitting (ISAS) iteration for solving the linear operator equation (128). The convergence analysis of the incomplete version of SAS iteration is given in [13].

3.5.2 Minimal Residual Iteration Method.

Minimal residual algorithm

The following fixed point equation can be derived from the definition of the SAS iteration (133):

$$\begin{cases} (\alpha I + S)u_1 = (\alpha I - A)u_2 + q \\ (\alpha I + A)u_2 = (\alpha I - S)u_1 + q \end{cases}. \tag{141}$$

In the operator form, the system (141) reads

$$\mathbf{T}(\alpha)\mathbf{u} = \mathbf{q}, \tag{142}$$

where the matrix of operators $\mathbf{T}(\alpha)$ and the vector functions $\mathbf{u}$ and $\mathbf{q}$ are defined as follows

$$\mathbf{T}(\alpha) = \begin{pmatrix} (\alpha I + S) & -(\alpha I - A) \\ -(\alpha I - S) & (\alpha I + A) \end{pmatrix}, \quad \mathbf{u} = \begin{pmatrix} u_1 \\ u_2 \end{pmatrix} \text{ and } \mathbf{q} = \begin{pmatrix} q \\ q \end{pmatrix}. \tag{143}$$

Let α be a positive constant. Since A is m-accretive and S is positive definite, for any $\alpha > 0$, the solution of problem (142) exists and is unique in $\mathcal{D}(T) \times \mathcal{D}(T)$ (see Theorem 3.3.2). It also holds that $u^* \in \mathcal{D}(T)$ is solution to (128) if and only if $\mathbf{u}^* = \begin{pmatrix} u_1^* \\ u_2^* \end{pmatrix} \in \mathcal{D}(T) \times \mathcal{D}(T)$ is solution to (142) (see Theorem 3.3.3). Thus problem (128) and problem (142) are equivalent.

Let $\mathbf{P}(\alpha)$ be the matrix operator defined in $\mathcal{D}(T) \times \mathcal{D}(T)$ by:

$$\mathbf{P}(\alpha) = \begin{pmatrix} (\alpha I + S) & 0 \\ 0 & (\alpha I + A) \end{pmatrix}. \tag{144}$$

The preconditioning of the system (142) from the right by $[\mathbf{P}(\alpha)]^{-1}$ leads to the following system

$$\mathbf{T}_1(\alpha)\mathbf{u} = \mathbf{q} \tag{145}$$

where the matrix of operator $\mathbf{T}_1(\alpha)$ reads

$$\mathbf{T}_1(\alpha) = \begin{pmatrix} I & -A_1(\alpha) \\ -S_1(\alpha) & I \end{pmatrix}, \tag{146}$$

The solution $\mathbf{v}^*$ of problem (142) reads

$$\mathbf{v}^* = [\mathbf{P}(\alpha)]^{-1}\mathbf{u}^*, \tag{147}$$

where $\mathbf{u}^*$ is solution of (145).

Since all the operators of the matrix $\mathbf{T}_1(\alpha)$ are bounded on H, $\mathbf{T}_1(\alpha)$ is bounded on $H \times H$.

We consider in $H \times H$ the inner product $\langle , \rangle$ defined for $\mathbf{u} = \begin{pmatrix} u_1 \\ u_2 \end{pmatrix}, \mathbf{v} = \begin{pmatrix} v_1 \\ v_2 \end{pmatrix} \in H \times H$ by

$$\langle \mathbf{u}, \mathbf{v} \rangle = (u_1, v_1) + (u_2, v_2), \tag{148}$$

and the associated norm

$$\||\mathbf{u}\|| = \left(\|u_1\|^2 + \|u_2\|^2 \right)^{\frac{1}{2}}. \tag{149}$$

The minimal residual iteration method for the solution of problem (145) minimizes the functional $\varepsilon(\mathbf{u}) = \||\mathbf{q} - \mathbf{T}_1(\alpha)\mathbf{u}\||^2$ as follows (see [6, 7, 50, 80]):

101

Algorithm 3.5.1. *Minimal Residual Algorithm*

Let $\mathbf{u}^{(0)} \in H \times H$, $\mathbf{r}^{(0)} = \mathbf{q} - \mathbf{T}_1(\alpha)\mathbf{u}^{(0)}$, $\mathbf{p}^{(0)} = \mathbf{r}^{(0)}$, $\mathbf{x}^{(0)} = \mathbf{T}_1(\alpha)\mathbf{p}^{(0)}$.

`While` $\|\|\mathbf{r}^{(n)}\|\| > \lambda_k$ `do`

`begin`

$$\rho^{(n)} = \langle \mathbf{r}^{(n)}, \mathbf{x}^{(n)}\rangle / \langle \mathbf{x}^{(n)}, \mathbf{x}^{(n)}\rangle;$$

$$\mathbf{u}^{(n+1)} = \mathbf{u}^{(n)} + \rho^{(n)}\mathbf{p}^{(n)};$$

$$\mathbf{r}^{(n+1)} = \mathbf{r}^{(n)} - \rho^{(n)}\mathbf{x}^{(n)};$$

$$\theta^{(n+1)} = -\langle \mathbf{T}_1(\alpha)\mathbf{r}^{(n+1)}, \mathbf{x}^{(n)}\rangle / \langle \mathbf{x}^{(n)}, \mathbf{x}^{(n)}\rangle;$$

$$\mathbf{p}^{(n+1)} = \mathbf{r}^{(n+1)} + \theta^{(n+1)}\mathbf{p}^{(n)};$$

$$\mathbf{x}^{(n+1)} = \mathbf{T}_1(\alpha)\mathbf{r}^{(n+1)} + \theta^{(n+1)}\mathbf{x}^{(n)};$$

`end.`

In the previous algorithm, we have to make clear how the product $\mathbf{T}_1(\alpha)$ times a vector is computed, since $T_1(\alpha)$ contains some inverse operator. Let $\mathbf{u} = (u_1, u_2)^t \in H \times H$. The components of $\mathbf{T}_1(\alpha)\mathbf{u} = (v_1, v_2)^t$ verify

$$\begin{cases} v_1 &= u_1 - (A - \alpha I)\varphi_2 \\ v_2 &= u_2 - (S - \alpha I)\varphi_1 \end{cases}, \tag{150}$$

where $\varphi_1 \in H$ satisfies the integral equation

$$(S + \alpha I)\varphi_1 = u_1 \tag{151}$$

and $\varphi_2 \in D(T)$ satisfies the differential equation

$$(A + \alpha I)\varphi_2 = u_2 . \tag{152}$$

Once φ_1 and φ_2 are calculated, the components of $\mathbf{T}_1(\alpha)\mathbf{u}$ in (153) are easily computed. The integral equation (155) and the differential equation (156) can be solved numerically.

Convergence results

From the analysis of the minimal residual algorithm (see [50]), the following estimate on the residual holds:

Theorem 3.5.1. *Given an initial guess* $\mathbf{u}^{(0)}$. *If the functions* $\mathbf{u}^{(k)}$ *(k > 0) are computed by the minimal residual algorithm, then*

$$\varepsilon(\mathbf{u}^{(k+1)}) \leq \lambda(\alpha, k)\varepsilon(\mathbf{u}^{(k)}), \tag{153}$$

where

$$\lambda(\alpha, k) = \left(1 - \frac{\langle \mathbf{r}^{(k)}, \mathbf{T}_1(\alpha)\mathbf{r}^{(k)}\rangle}{\langle \mathbf{r}^{(k)}, \mathbf{r}^{(k)}\rangle} \frac{\langle \mathbf{r}^{(k)}, \mathbf{T}_1(\alpha)\mathbf{r}^{(k)}\rangle}{\langle \mathbf{T}_1(\alpha)\mathbf{r}^{(k)}, \mathbf{T}_1(\alpha)\mathbf{r}^{(k)}\rangle}\right). \tag{154}$$

The convergence of the minimal residual method is guaranteed, if $\lambda(\alpha, k) < 1$.

Let α be a positive constant. The following properties are characteristic of the operator $\mathbf{T}_1(\alpha)$:

Theorem 3.5.2. *For all* $\mathbf{u} \in H \times H$, *the following inequalities hold true:*

$$\langle \mathbf{T}_1(\alpha)\mathbf{u}, \mathbf{u}\rangle \geq \frac{1 - \beta(\alpha)}{2} |||\mathbf{u}|||^2; \tag{155}$$

$$\langle \mathbf{T}_1(\alpha)\mathbf{u}, \mathbf{u}\rangle \geq \frac{1}{2}\langle \mathbf{T}_1(\alpha)\mathbf{u}, \mathbf{T}_1(\alpha)\mathbf{u}\rangle. \tag{156}$$

Proof. Let $\mathbf{u} = \begin{pmatrix} u_1 \\ u_2 \end{pmatrix}$. We have

$$
\begin{aligned}
\langle \mathbf{T}_1(\alpha)\mathbf{u}, \mathbf{u}\rangle &= \|u_1\|^2 + \|u_2\|^2 - (A_1(\alpha)u_2, u_1) - (S_1(\alpha)u_1, u_2) \\
&\geq \|u_1\|^2 + \|u_2\|^2 - (\|A_1(\alpha)\| + \|S_1(\alpha)\|)\|u_1\|\|u_2\| \\
&\geq \||\mathbf{u}\||^2 - (1 + \beta(\alpha))\|u_1\|\|u_2\| \\
&\geq \frac{1-\beta(\alpha)}{2}\||\mathbf{u}\||^2;
\end{aligned}
$$

and

$$
\begin{aligned}
\langle \mathbf{T}_1(\alpha)\mathbf{u}, \mathbf{u}\rangle - \tfrac{1}{2}\langle \mathbf{T}_1(\alpha)\mathbf{u}, \mathbf{T}_1(\alpha)\mathbf{u}\rangle &= \tfrac{1}{2}\left(\|u_1\|^2 + \|u_2\|^2\right) \\
&\quad - \tfrac{1}{2}\left(\|S_1(\alpha)u_1\|^2 + \|A_1(\alpha)u_2\|^2\right) \\
&\geq \tfrac{1}{2}(1 - \beta^2(\alpha))\|u_1\|^2 \\
&\geq 0.
\end{aligned}
$$

$\square$

Theorem 3.5.3. *Convergence results.*

Let α be a positive constant. Given an initial guess $\mathbf{u}^{(0)} \in H \times H$, if the sequence $\{\mathbf{u}^{(k)}\}_{k \geq 0}$ is obtained by the minimal residual algorithm, then the following error estimations hold

$$\varepsilon(\mathbf{u}^{(k+1)}) \leq \frac{3 + \beta(\alpha)}{4}\varepsilon(\mathbf{u}^{(k)}), \tag{157}$$

$$\||\mathbf{u}^{(k+1)} - \mathbf{u}^*\|| \leq \frac{2}{1 - \beta(\alpha)}\varepsilon(\mathbf{u}^{(k+1)})^{\frac{1}{2}}. \tag{158}$$

where $\mathbf{u}^*$ is the exact solution of problem (145). Thus $\left\{\mathbf{u}^{(k)}\right\}_{k\geq 0}$ converges toward $\mathbf{u}^*$.

Proof. We deduce from inequalities (155) and (156) that:

$$\lambda(\alpha, k) \leq \frac{1 - \beta(\alpha)}{4}, \quad k \geq 0.$$

It then follows from (153) that $\varepsilon(\mathbf{u}^{(k+1)}) \leq \frac{3+\beta(\alpha)}{4}\varepsilon(\mathbf{u}^{(k)})$.

Substituting $\mathbf{u}$ in (155) by $(\mathbf{u}^{(k+1)} - \mathbf{u}^*)$ yields

$$\begin{aligned}
|||\mathbf{u}^{(k+1)} - \mathbf{u}^*|||^2 &\leq \tfrac{2}{1-\beta(\alpha)}\langle \mathbf{T}_1(\alpha)\mathbf{u}^{(k+1)} - \mathbf{q}, \mathbf{u}^{(k+1)} - \mathbf{u}^*\rangle \\
&\leq \tfrac{2}{1-\beta(\alpha)}\left(\varepsilon(u^{(k+1)})\right)^{\frac{1}{2}} \cdot |||\mathbf{u}^{(k+1)} - \mathbf{u}^*|||,
\end{aligned}$$

and the estimation (158) then follows.

$\square$

3.5.3 Numerical Results

We apply the preceding minimal residual algorithm for solving neutron transport equation in 2-D geometry.

The neutron transport equation

The single group steady state first order neutron transport equation in 2-D cartesian geometry verified by the neutron flux $u(r,\omega) : Q := R \times B \to \mathbb{R}^+$, reads:

$$\begin{cases} Tu(r,\mathbf{\Omega}) := Au(r,\mathbf{\Omega}) + \Sigma u(r,\mathbf{\Omega}) = Ku(r,\mathbf{\Omega}) + q(r,\mathbf{\Omega}), & (r,\mathbf{\Omega}) \in Q \\ u(r,\mathbf{\Omega}) \in D(T) \end{cases} \tag{159}$$

The operators A, Σ and K are defined by:

$$Au = \mathbf{\Omega}\nabla_r u; \quad \Sigma u = \sigma(r)u; \quad Ku = \int_B \kappa(r,\mathbf{\Omega},\mathbf{\Omega}')u(r,\mathbf{\Omega}')d\mathbf{\Omega}' \tag{160}$$

and,

$$D(T) = \{u \in L^2(Q) : \mathbf{\Omega}\nabla_r u \in L^2(Q) \text{ and } u_{|\Gamma_-} = 0\}, \tag{161}$$

where $\Gamma_- = \{r \in \partial R \times B : \mu n_x + \eta n_y < 0\}$ ($n = (n_x, n_y)$ being the outer normal to ∂R), $r = (x,y)$, $\mathbf{\Omega} = (\mu, \eta)$, $R =]0,1[\times]0,1[$ and $B = \{\mathbf{\Omega} \in \mathbb{R}^2 : |\mathbf{\Omega}| < 1\}$. The function $\sigma(r)$ represents the total cross section and $\kappa(r,\mathbf{\Omega},\mathbf{\Omega}')$ is a non negative kernel describing the scattering of particles. The function q is the non negative source term.

104

The operator A is m-accretive (see [34]). In the following, it is assumed that:

(a1) $\sigma \in L^\infty(Q)$ and $\exists \sigma_0 > 0$ such that $\sigma(r) \geq \sigma_0$ a.e on Q;

(a2) $\kappa(r, \Omega, \Omega') = \kappa(x, \Omega', \Omega)$ and κ is non negative;

(a3) $\exists c \in [0, 1), \int_B \kappa(r, \Omega, \Omega') d\Omega' \leq \sigma_0 c$ a.e on Q.

From assumptions $(a1) - (a3)$ made on σ and κ, follows that operator $S = \sigma I - K$ is self-adjoint and positive definite (see [34]). Thus T is positive definite and it follows the existence and uniqueness of the solution of problem (159). Moreover, T admits a self-adjoint positive definite and m-accretive splitting (SAS) which yields the SAS iteration method. The SAS iteration method and the preceding minimal residual method for the solution of equation (159) converge. The equation (159) is known to be near singular when $c \approx 1$ (see [34]).

In the case of isotropic scattering where the integral operator is defined by

$$K\psi = \sigma_s(x) Pu, \tag{162}$$

with

$$Pu = \frac{1}{\pi} \int_B u(x, \Omega') d\Omega',$$

the inverse of the operator $(\alpha I + S)$ is given by (see [13]):

$$(\alpha I + S)^{-1} = \frac{1}{\sigma(x) - \sigma_s(x) + \alpha} P + \frac{1}{\sigma(x) + \alpha}(I - P). \tag{163}$$

Therefore, the linear integral equation (155) can be solved explicitly. Moreover, the second component v_2 in (153) can be calculated as follows:

$$v_2 = u_2 - P_1 u_1, \tag{164}$$

where

$$P_1 = \left(\frac{\alpha - \sigma - \sigma_s}{\alpha + \sigma - \sigma_s} - \frac{\alpha - \sigma}{\alpha + \sigma} \right) P + \frac{\alpha - \sigma}{\alpha + \sigma} I. \tag{165}$$

Here, σ_s and $\sigma_a = \sigma - \sigma_s$ denote the scattering and the absorption cross sections respectively. For the physical interest, the scattering ratio and the optical coefficient are respectively defined as follows:

$$\gamma = \max_{x \in R} \left(\frac{\sigma_s(x)}{\sigma_s(x) + \sigma_a(x)} \right) \quad \text{and} \quad \nu = \min_{x \in R} (\sigma_s(x) + \sigma_a(x)) \operatorname{diam}(R), \tag{166}$$

where $\operatorname{diam}(R)$ denotes the diameter of domain R. The values $\gamma = 1$ and $\nu >> 1$ ($\sigma_a >> 1$) correspond to pure scattering and optically thick domains respectively, and represent two extreme situations in computational transport where conventional discretization methods such as piecewise linear finite elements using Galerkin formulation [57], classical finite difference scheme [49] and upwind difference scheme [56] yield inaccurate solutions unless the spatial grid is very fine. As mentioned in [66, 55], as $\sigma_t = \sigma_a + \sigma_s$ tends to infinity and γ tends to 1, the problem becomes singularly perturbed. Therefore discrete approximation to the transport problem using these methods will have operators with condition numbers on the order of at least σ_t^2 regardless of the mesh size [55].

Discretization and numerical results

The discretization is carried out by a DSN scheme (see [13, 34]) consisting of using a finite set of L discrete angular directions $\Omega_L = \{\Omega_i = (\mu_i, \eta_i)\}_{i=1}^{i=L} \subset B$, which have nonzero components and are symmetric about the origin for the angular approximation and, a difference method based on control volume approach and cell averaging for the spatial approximation. The numerical grid is defined by:

$$R_h = \{(x_i, y_j), 0 \le i \le N, 0 \le i \le M\}, \tag{167}$$

where $x_0 = 0$, $x_i = x_{i-1} + (\Delta x)_i$, $x_N = 1$, $y_0 = 0$, $y_j = y_{j-1} + (\Delta y)_j$, $y_M = 1$ and $h = \max_{ij}((\Delta x)_i, (\Delta y)_j)$. The cell center grid points are defined as: $x_{i+\frac{1}{2}} = \dfrac{x_{i+1} - x_i}{2}$, $y_{j+\frac{1}{2}} = \dfrac{y_{j+1} - y_j}{2}$, $(\Delta x)_{i+\frac{1}{2}} = x_{i+1} - x_i$ and $(\Delta y)_{j+\frac{1}{2}} = y_{j+1} - y_j$.

For the numerical results, we took particular data for which an exact solution of problem (159) is known in each case. For the iterative methods tested here, the iterations are stopped when the relative error $\frac{\|U - U_{exate}\|_2}{\|U_{exate}\|_2}$ is less than a prescribed $\epsilon > 0$.

For $x = (x_1, x_2) \in R$ and $\Omega = (\mu, \eta) \in B$, we set $\sigma(x) = \sigma$, $\kappa(x, \Omega, \Omega') = \frac{\sigma c}{\pi}$ and

$$q(x, \mu) = \begin{cases} \mu x_2 + \eta x_1 + \sigma x_1 x_2 - \frac{\sigma c}{4}, & \mu > 0, \eta > 0; \\ -\mu x_2 + \eta(1 - x_1) + \sigma(1 - x_1)x_2 - \frac{\sigma c}{4}, & \mu < 0, \eta > 0; \\ -\mu(1 - x_2) - \eta(1 - x_1) + \sigma(1 - x_1)(1 - x_2) - \frac{\sigma c}{4}, & \mu < 0, \eta < 0; \\ \mu(1 - x_2) - \eta x_1 + \sigma x_1(1 - x_2) - \frac{\sigma c}{4}, & \mu > 0, \eta < 0. \end{cases}$$

The exact solution of this test problem is given by:

$$\psi(x, \mu) = \begin{cases} x_1 x_2, \ \mu > 0, \eta > 0; & (1 - x_1)x_2, \mu < 0, \eta > 0; \\ (1 - x_1)(1 - x_2), \ \mu < 0, \eta < 0; & x_1(1 - x_2), \mu > 0, \eta < 0. \end{cases}$$

In this problem, σc is the scattering cross section and $\sigma_a = \sigma(1 - c)$ represents the absorption cross section of neutrons in R. The parameter c is the scattering ratio and σ is the optical coefficient.

For the numerical test, we take $\Delta x = \Delta y = \frac{1}{10}$ and L=100. We study the behavior of the Minimal residual method with respect to parameter σ, c and α. For the example problem, the theoretical optimal parameter which minimizes the bound $\beta(\alpha)$ is $\alpha_t = \sigma_a = \sigma(1-c)$. It is observed in [13] that, for fixed c and σ, numerical optimal value of α can be localized in the interval $[\sigma(1-c), \sigma(1-c/2)]$. The value

$$\alpha* = \begin{cases} \sigma(1-19c/32), & 0 < c < 0.9 \\ \sigma(1-23c/32), & 0.9 \leq c < 0.97 \\ 1, & 0.97 \leq c \leq 1 \end{cases}$$

yielded good convergence results for the SAS iteration applied to the example problem as compare to the standard source ietration method, a spatial multigrid algrithm and some Krylov subspace methods such as GMRES and BiCGStab iterative algorithms (see [13]). The minimal residual iteration method was applied to the example problem for several values of c and σ using the parameter α_t. We observed the convergence of the method. It is also observed for some values of σ_a and c ($\sigma_a \geq 4$ and $c > 0.5$), the convergence of the minimal residual seems to be faster as compare to the SAS method using α_t. This results also holds for large value of σ when using the SAS method with $\alpha*$. For the other cases, we used a Gauss-Seidel preconditioned version of the minimal residual algorithm.

We presented comparative numerical results (number of iterations) of the previous minimal residual algorithm (Minres), the Minres with Gauss-Seidel preconditioning (GS-Minres), the SOR acceleration of SAS with relaxation parameter $w = 0.93$, setting $\alpha = \alpha_t$ (we set $\alpha = \alpha_t + c$ for values of c close to 1) and the SAS iterations using $\alpha = \alpha_t$ and $\alpha = \alpha*$. There are two sets of tests: one at fixed σ and another at fixed c. As shown by Figure 29 to Figure 34, all the methods converge. At fixed $\sigma = 100$ and $\sigma = 1000$, we compare the $c-$dependence of the iterative methods used here. The Figure 29 and figure 30 plot the number of iterations of methods as function of $c \in [0.1, 0.99]$ for $\sigma = 100$ and $\sigma = 1000$ respectively. We observe in Figure 29 ($\sigma = 100$) that the Minres method is faster than the SAS(σ_t) for $0.5 < c \leq 0.96$ and is slower than SAS($\alpha*$) $0.1 \leq c \leq 0.99$. For c close to 1 ($c \in [0.98, 0.99999]$), it can be observed in Figure 31 that, for $\sigma = 100$ the Minres method is faster than the SOR method for $c > 0.9975$. Setting $\sigma = 1000$, we observed that in Figure 32 the Minres method is faster than the SAS(α_t) except for

107

	SAS	SOR	Minres	GS-Minres
$\sigma = 50$	203	286	248	136
$\sigma = 100$	292	436	361	192
$\sigma = 1000$	487	850	643	334

Table 4: Number of iteration of SAS, SOR, Minres and GS-Minres for $c = 1$ ($\alpha = 1$)

$c = 0.999$, and SAS($\alpha*$) for $0.98 \leq c \leq 0.995$. It can be seen that, for this set of test, the GS-Minres method is efficient than all the methods tested here.

The σ−dependence of iterative methods at fixed $c = 0.98$ is plotted by Figure 33 and Figure 34 for $\sigma \in [1, 500]$ and for large values of σ respectively. It can be observed that the Minres algorithm is faster than SAS(α_t), SAS($\alpha*$) and SOR methods for $\sigma > 150$, $\sigma > 200$ and $\sigma > e^7$ respectively and it is comparable to GS-Minres method for $\sigma = e^{10}$. The Table 4 presents the number of iteration of the methods tested here at $c = 1$ for $\sigma \in \{50, 100, 1000\}$. It can be observed tha, the number of iteration increase with the value of σ. The GS-Minres presents good convergence results as compare to other method. Finally We notice that the GS-Minres method is efficient than all the methods for this set of tests.

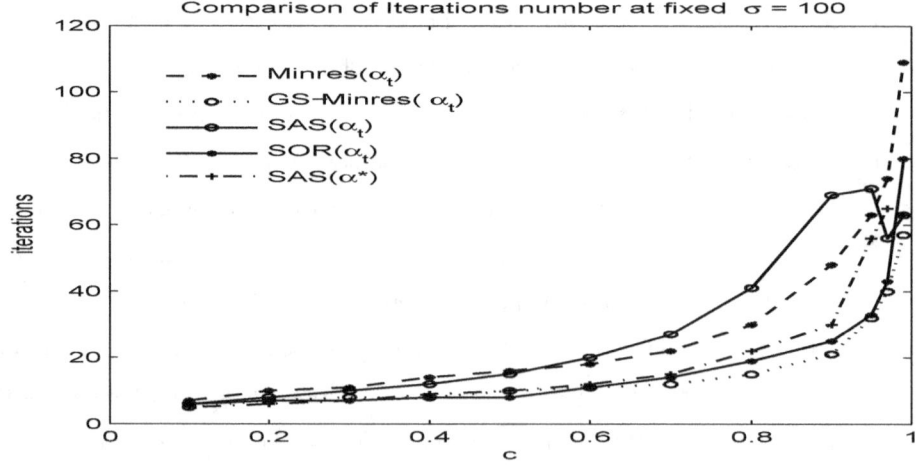

Figure 29: Comparison of methods at fixed $\sigma = 100$, for $c \in [0.1, 0.99](\epsilon = 1E - 08)$.

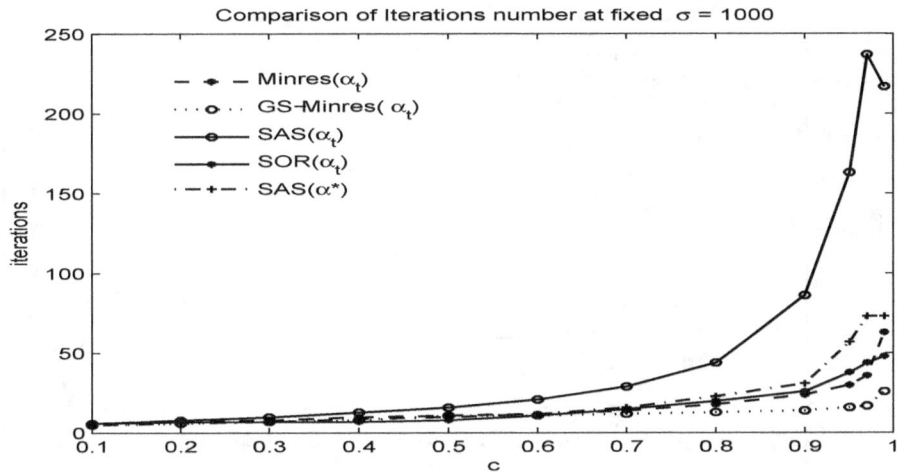

Figure 30: Comparison of methods at fixed $\sigma = 1000$, for $c \in [0.1, 0.99](\epsilon = 1E - 08)$.

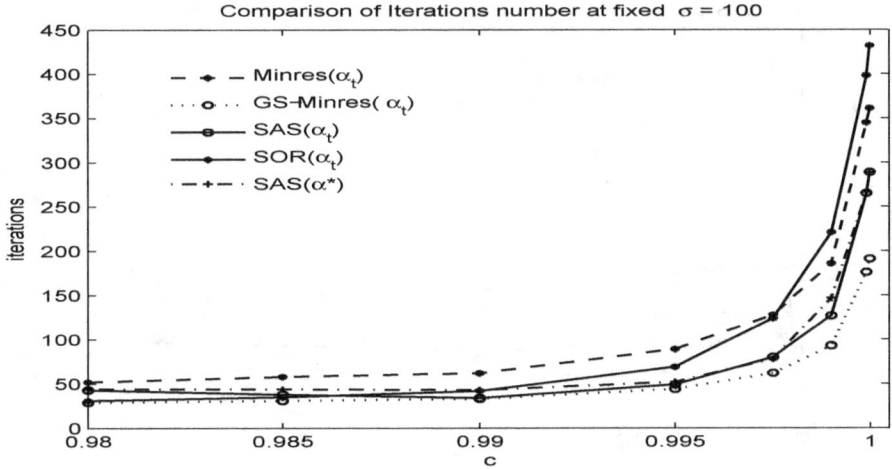

Figure 31: Comparison of methods at fixed $\sigma = 100$, for $c \in [0.98, 0.99999](\epsilon = 1E - 06)$.

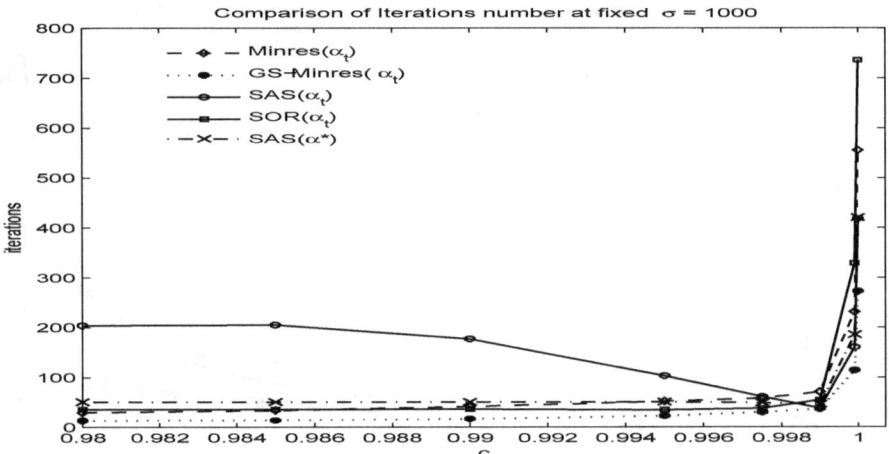

Figure 32: Comparison of methods at fixed $\sigma = 1000$, for $c \in [0.98, 0.99999](\epsilon = 1E-06)$.

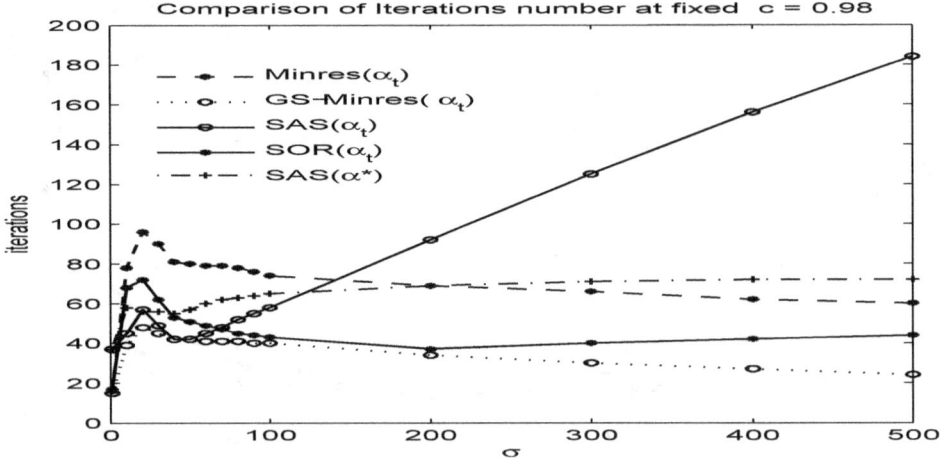

Figure 33: Comparison of methods at fixed $c = 0.98$, for $\sigma \in [1, 500](\epsilon = 1E - 08)$.

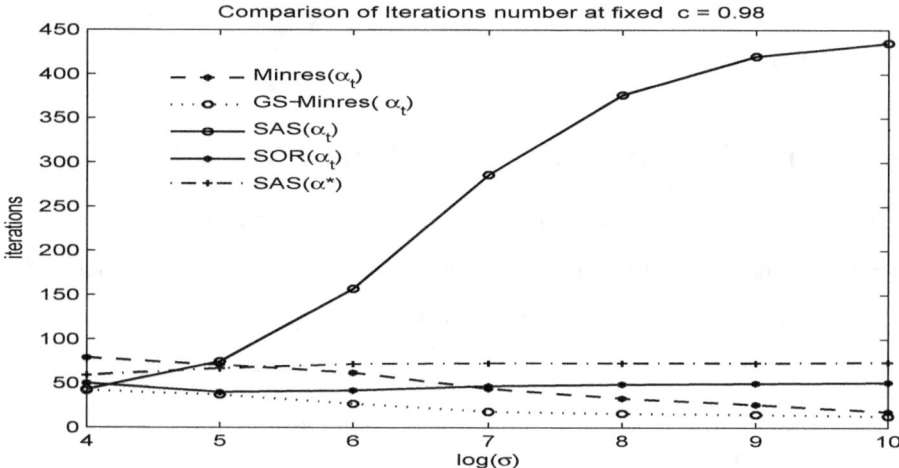

Figure 34: Comparison of methods at fixed $c = 0.98$, for large value of σ ($\sigma \in [e^4, e^{10}]$ and $\epsilon = 1E - 08$).

3.6 A Preconditioned Minimal Residual Solver for a Class of Linear Operator Equations

In the previous section, an infinite dimensional adaptation of a minimal residual algorithm has been applied for solving a fixed point problem in $H \times H$ derived from the Self-adjoint and m-accretive splitting iteration method. The theoretical analysis shows the convergence of the method and an upper bound for the rate of residual decreasing has be derived. The method and its Gauss-Seidel preconditioned version have been successfully applied to a a sample neutron transport problem in 2-D geometry. In this section, we introduced preconditioned versions of this minimal residual method using symmetric Gauss-Seidel and polynomial preconditioning. Theoretical analysis shows that the methods converge unconditionally and upper bounds of the rate of residual decreasing which depend solely on the spectrum of the self-adjoint part of the operator are derived.

The remainder of this section is structured as follows: in subsection 3.6.1 we give the description and the convergence properties of the SAS and minimal residual iteration method. The symmetric Gauss-Seidel preconditioning and polynomial preconditioning are presented in subsection 3.6.2 and the convergence analysis of the preconditioned version of the minimal residual is considered in subsection 3.6.3. The application of the method to 2-D neutron transport equation and the numerical illustration are presented in subsection 3.6.4.

3.6.1 The SAS and Minimal Residual Iteration Method

Let us consider a Hilbert space H with inner product $(.,.)$ and norm $\|.\|$ and let T be a linear operator on H with domain $\mathcal{D}(T)$ and range $\mathcal{R}(T) = H$. We denote by I, the identity operator. Suppose that we need to solve in $\mathcal{D}(T)$, the following problem

$$Tu = q, \tag{168}$$

where $q \in H$ is given and $u \in \mathcal{D}(T)$ is the unknown.

We assume that the operator T admits the following splitting (see [13, 16]):

$$T = S + A, \tag{169}$$

where S is a bounded self-adjoint positive definite operator and A is a m-Accretive operator. Therefore, the operator T is positive definite and equation (168) admits a unique solution in H.

Let α be a positive constant. The following two-step splitting is obtained from (169):

$$\begin{cases} T = (\alpha I + S) - (\alpha I - A) \\ T = (\alpha I + A) - (\alpha I - S) \end{cases}. \tag{170}$$

which leads to the following Self-adjoint and m-Accretive Splitting (SAS) iteration method (see [13]): Given an initial guess $u^{(0)} \in D(T)$, for $k = 0, 1, \ldots$ until $\{u^{(k)}\}$ converges, calculate

$$\begin{cases} (\alpha I + S)u^{(k+\frac{1}{2})} = (\alpha I - A)u^{(k)} + q \\ (\alpha I + A)u^{(k+1)} = (\alpha I - S)u^{(k+\frac{1}{2})} + q \end{cases}. \tag{171}$$

The exact solution u^* of the problem (168) verifies (see [13, 16]) :

$$\|u^{(k+1)} - u^*\|_{A(\alpha)} \leq \beta(\alpha)\|u^{(k)} - u^*\|_{A(\alpha)}, \tag{172}$$

where $\|.\|_{A(\alpha)}$ is a norm defined on $\mathcal{D}(T)$ by

$$\|u\|_{A(\alpha)} = \|(\alpha I + A)u\|, \tag{173}$$

and

$$\beta(\alpha) = \sup_{\lambda \in \sigma(S)} \left| \frac{\alpha - \lambda}{\alpha + \lambda} \right|, \tag{174}$$

with $\sigma(S)$ denoting the spectrum of S. It holds from the positivity of α and λ that $\beta(\alpha) < 1$. Thus the SAS iteration (171) converges unconditionally to the solution of (168) with respect to norm $\|.\|_{\alpha}$. Since for $u \in D(T)$, we have $\alpha\|u\| \leq \|u\|_{A(\alpha)}$, the convergence of the SAS iteration with respect to the norm $\|.\|$ follows. The theoretical optimal parameter α^* for the bound $\beta(\alpha)$ is $\alpha^* = \sqrt{\lambda_{min}\lambda_{max}}$ with λ_{min} and λ_{max} denoting respectively the lower and upper bounds of $\sigma(S)$ (see [21, 59]). The convergence analysis of the incomplete version of SAS iteration where each subproblem of (171) is solved approximately, is given in [13].

The following fixed point equation can be derived from the definition of the SAS iteration (171):

$$\begin{cases} (\alpha I + S)u_1 = (\alpha I - A)u_2 + q \\ (\alpha I + A)u_2 = (\alpha I - S)u_1 + q \end{cases}, \tag{175}$$

which is equivalent to the following linear operator equation in $\mathcal{D}(T) \times \mathcal{D}(T)$:

$$\mathbf{T}(\alpha)\mathbf{u} = \mathbf{q}, \tag{176}$$

where the matrix of operators $\mathbf{T}(\alpha)$ and the vector functions $\mathbf{u}$ and $\mathbf{q}$ are defined by (143).

The solution of the linear operator equation (176) is given by $\mathbf{u}^* = \begin{pmatrix} u^* \\ u^* \end{pmatrix}$ where u^* is the solution of (168) (see [12]). Then it follows that problems (168) and (176) are equivalent.

The preconditioning of the system (176) from the right by $[\mathbf{P}(\alpha)]^{-1}$, where $\mathbf{P}(\alpha)$ is given by (144), leads to the following system

$$\mathbf{T}_1(\alpha)\mathbf{v} = \mathbf{q} \tag{177}$$

where

$$\mathbf{v} = [\mathbf{P}(\alpha)]^{-1}\mathbf{u} \tag{178}$$

and $\mathbf{T}_1(\alpha)$ defined by (146).

Let $\|\|.\|\|$ be the norm defined on $H \times H$ by (149). The minimal residual iteration method for the solution of problem (177) results from the minimization of the residual functional $\varepsilon(\mathbf{u}) = \|\|\mathbf{q} - \mathbf{T}_1(\alpha)\mathbf{u}\|\|^2$, using the Algorithm 3.5.1 (see [6, 7, 50, 80]).

The following estimate on the residual holds from the analysis of this minimal residual algorithm (see [50]): Given an initial guess $\mathbf{u}^{(0)}$. If the functions $\mathbf{u}^{(k)}$ $(k > 0)$ are computed by the minimal residual algorithm, then

$$\varepsilon(\mathbf{u}^{(k+1)}) \leq \lambda_1(\alpha, k)\varepsilon(\mathbf{u}^{(k)}), \tag{179}$$

113

where

$$\lambda_1(\alpha, k) = \left(1 - \frac{\langle \mathbf{r}^{(k)}, \mathbf{T}_1(\alpha)\mathbf{r}^{(k)} \rangle}{\langle \mathbf{r}^{(k)}, \mathbf{r}^{(k)} \rangle} \frac{\langle \mathbf{r}^{(k)}, \mathbf{T}_1(\alpha)\mathbf{r}^{(k)} \rangle}{\langle \mathbf{T}_1(\alpha)\mathbf{r}^{(k)}, \mathbf{T}_1(\alpha)\mathbf{r}^{(k)} \rangle} \right). \tag{180}$$

The convergence results of the minimal residual method for the solution of (177) are given by the Theorem 3.5.3.

3.6.2 Symmetric Gauss-Seidel and polynomial preconditioning

We present in this section two split type preconditioning strategies of the problem (177). The first one consist of symmetric Gauss-Seidel preconditioning and the second one is a couples symmetric Gauss-Seidel and polynomial preconditioning. Let us consider in $H \times H$ the following operators:

$$\mathbf{M}_1(\alpha) = \begin{pmatrix} I & 0 \\ -S_1(\alpha) & I \end{pmatrix}, \quad \mathbf{M}_2(\alpha) = \begin{pmatrix} I & -A_1(\alpha) \\ 0 & I \end{pmatrix}. \tag{181}$$

The operators $\mathbf{M}_1(\alpha)$ and $\mathbf{M}_2(\alpha)$ are bounded and have bounded inverses defined by:

$$\mathbf{M}_1^{-1}(\alpha) = \begin{pmatrix} I & 0 \\ S_1(\alpha) & I \end{pmatrix}, \quad \mathbf{M}_2^{-1}(\alpha) = \begin{pmatrix} I & A_1(\alpha) \\ 0 & I \end{pmatrix}. \tag{182}$$

The symmetric Gauss-Seidel preconditioner of the problem (177) is defined by:

$$\mathbf{M}_{SGS}(\alpha) = \mathbf{M}_1(\alpha)\mathbf{M}_2(\alpha). \tag{183}$$

The split preconditioning of (177) using $\mathbf{M}_{SGS}$ leads to the following equivalent problem:

$$\mathbf{T}_2(\alpha)\mathbf{v} = \mathbf{q}_2(\alpha), \tag{184}$$

$$\mathbf{u} = \mathbf{M}_2^{-1}(\alpha)\mathbf{v}, \tag{185}$$

where

$$\mathbf{T}_2(\alpha) = \mathbf{M}_1^{-1}(\alpha)\mathbf{T}_1(\alpha)\mathbf{M}_2^{-1}(\alpha) = \begin{pmatrix} I & 0 \\ 0 & I - M(\alpha) \end{pmatrix}, \quad \mathbf{q}_2(\alpha) = \mathbf{M}_1^{-1}(\alpha)\mathbf{q}_1,$$

with $M(\alpha) = S_1(\alpha)A_1(\alpha)$.

The operator of equation (184) can be written as

$$\mathbf{T}_2(\alpha) = \mathbf{I} - \mathbf{M}(\alpha), \tag{186}$$

114

where $\mathbf{I}$ denotes the identity operator in $H \times H$ and $\mathbf{M}(\alpha) = \begin{pmatrix} 0 & 0 \\ 0 & M(\alpha) \end{pmatrix}$. Since $\|M(\alpha)\| \leq \beta(\alpha)$, we have for $\mathbf{u} = (u_1, u_2)^t \in H \times H$,

$$\|\mathbf{M}(\alpha)\mathbf{u}\|\|^2 = \|M(\alpha)u_2\|^2 < \beta^2(\alpha)\|u_2\|^2 < \beta^2(\alpha)\|\|\mathbf{u}\|\|^2.$$

Thus

$$\|\mathbf{M}(\alpha)\|\| \leq \beta(\alpha) < 1, \tag{187}$$

and

$$\mathbf{T}_2^{-1}(\alpha) = (\mathbf{I} - \mathbf{M}(\alpha))^{-1} = \sum_{k=0}^{\infty} \mathbf{M}^k(\alpha). \tag{188}$$

Therefore, the operator $\mathbf{T}_2^{-1}(\alpha)$ can be approximated by the following truncated Neumann series:

$$\mathbf{P}_n(\alpha) = \sum_{k=0}^{n} \mathbf{M}^k(\alpha). \tag{189}$$

Setting

$$\mathbf{T}_2(\alpha, n) = \mathbf{P}_n(\alpha)\mathbf{T}_2(\alpha) = \mathbf{I} - \mathbf{M}^{n+1}(\alpha) \quad \text{and} \quad \mathbf{q}_2(\alpha, n) = \mathbf{P}_n(\alpha)\mathbf{q}_2(\alpha), \tag{190}$$

we obtain the following operator equation :

$$\mathbf{T}_2(\alpha, n)\mathbf{u} = \mathbf{q}_2(\alpha, n), \tag{191}$$

which is equivalent to (184).

We have

$$\mathbf{T}_2(\alpha, n) = \mathbf{C}_1^{-1}(\alpha)\mathbf{T}_1(\alpha)\mathbf{M}_2^{-1}(\alpha) \text{ and } \mathbf{q}_2 = \mathbf{C}_1^{-1}(\alpha)\mathbf{q}, \tag{192}$$

where $\mathbf{C}_1(\alpha) = \mathbf{P}_n^{-1}(\alpha)\mathbf{M}_1(\alpha)$. Thus equation (191) follows from a split preconditioning of equation (177), using $\mathbf{C}_1(\alpha)\mathbf{M}_2(\alpha)$ as a preconditioner. This can be seen as a couple symmetric Gauss-Seidel and polynomial preconditioning of equation (177). It can be noticed that equation (184) is a particular case of equation (191), when $n = 0$.

To apply the minimal residual algorithm for the solution of equation (184), we have to make clear how the product $\mathbf{T}_2(\alpha, n)$ times a vector is computed, since $\mathbf{T}_2(\alpha, n)$ contains some inverse operators. Let $\mathbf{u} = \begin{pmatrix} u_1 \\ u_2 \end{pmatrix} \in H \times H$. The components of $\mathbf{T}_2(\alpha, n)\mathbf{u} = \begin{pmatrix} v_1 \\ v_2 \end{pmatrix}$ verify $v_1 = u_1$ and $v_2 = u_2 - M^{n+1}(\alpha)u_2$. The main task consists in computing the product $M^{n+1}(\alpha)u_2$, which is obtained after $n + 1$ successive computations of products

115

of the form $M(\alpha)u = \varphi$. Let $u \in H$, we describe in the following how to compute $\varphi = M(\alpha)u$. We have

$$\begin{cases} \varphi_2 = (A - \alpha I)\varphi_1 \\ \varphi = (S - \alpha I)\varphi_3, \end{cases} \qquad (193)$$

where $\varphi_1 \in \mathcal{D}(T)$ satisfies the differential equation

$$(A + \alpha I)\varphi_1 \;=\; u \qquad (194)$$

and $\varphi_3 \in H$ satisfies the integral equation

$$(S + \alpha I)\varphi_3 \;=\; \varphi_2 \;. \qquad (195)$$

Once φ_1, φ_2 and φ_3 are calculated, the product φ is easily computed. The differential equation (194) and the integral equation (195) can be solved numerically.

3.6.3 Convergence analysis of the preconditioned minimal residual method

We present in this section the convergence results of the minimal residual algorithm of section 3.5.2 applied to (191).

The following properties are characteristic of the operator $\mathbf{T}_2(\alpha, n)$:

Theorem 3.6.1. *Let α be a positive constant. For all $\mathbf{u} \in H \times H$, the following inequalities hold true:*

$$\langle \mathbf{T}_{(}\alpha, n)\mathbf{u}, \mathbf{u} \rangle \;\geq\; (1 - \beta^{n+1}(\alpha))|||\mathbf{u}|||^2, \qquad (196)$$

$$\langle \mathbf{T}_2(\alpha, n)\mathbf{u}, \mathbf{u} \rangle \;\geq\; \frac{1}{2}\langle \mathbf{T}_2(\alpha, n)\mathbf{u}, \mathbf{T}_2(\alpha, n)\mathbf{u} \rangle. \qquad (197)$$

Proof. Let $\mathbf{u} = \begin{pmatrix} u_1 \\ u_2 \end{pmatrix}$. We have

$$\begin{aligned}
\langle \mathbf{T}_2(\alpha, n)\mathbf{u}, \mathbf{u} \rangle \;&=\; \|u_1\|^2 + \|u_2\|^2 - (M^{n+1}(\alpha)u_2, u_2) \\
&\geq\; \|u_1\|^2 + \|u_2\|^2 - \|M^{n+1}(\alpha)\|\|u_2\|^2 \\
&\geq\; |||\mathbf{u}|||^2 - \|M^{n+1}(\alpha)\||||\mathbf{u}|||^2 \\
&\geq\; (1 - \beta^{n+1}(\alpha))|||\mathbf{u}|||^2.
\end{aligned}$$

116

We also have

$$\langle \mathbf{T}_2(\alpha, n)\mathbf{u}, \mathbf{u}\rangle - \tfrac{1}{2}\langle \mathbf{T}_2(\alpha, n)\mathbf{u}, \mathbf{T}_2(\alpha, n)\mathbf{u}\rangle = \tfrac{1}{2}\|\|\mathbf{u}\|\|^2 - \tfrac{1}{2}\|M^{n+1}(\alpha)u_1\|^2$$
$$\geq \tfrac{(1-\beta^{2n+2}(\alpha))}{2}\|\|\mathbf{u}\|\|^2 \geq 0.$$

Thus the inequality (197) is satisfied. □

Theorem 3.6.2. *Convergence results.*

Let α be a positive constant. Given an initial guess $\mathbf{u}^{(0)} \in H \times H$, if the sequence $\{\mathbf{u}^{(k)}\}_{k\geq0}$ for the approximation of the solution u^* of (191) is obtained by the minimal residual algorithm, then the following error estimations hold

$$\varepsilon(\mathbf{u}^{(k+1)}) \leq \frac{1 + \beta^{n+1}(\alpha)}{2}\varepsilon(\mathbf{u}^{(k)}), \tag{198}$$

$$\|\|\mathbf{u}^{(k+1)} - \mathbf{u}^*\|\| \leq \frac{1}{1 - \beta^{n+1}(\alpha)}\varepsilon(\mathbf{u}^{(k+1)})^{\frac{1}{2}}. \tag{199}$$

where $\mathbf{u}^*$ is the exact solution of problem (191). Thus $\{\mathbf{u}^{(k)}\}_{k\geq0}$ converges toward $\mathbf{u}^*$.

Proof. Substituting $\mathbf{T}_1(\alpha)$ in (180) by $\mathbf{T}_2(\alpha, n)$, we deduce from inequalities (196) and (197) the following bound for the rate of residual decreasing:

$$\lambda_3(\alpha, k) \leq \frac{1 + \beta^{n+1}(\alpha)}{2}, \quad k \geq 0$$

and the inequality (198) follows from (179).

Substituting $\mathbf{u}$ in (196) by $(\mathbf{u}^{(k+1)} - \mathbf{u}^*)$ yields

$$\|\|\mathbf{u}^{(k+1)} - \mathbf{u}^*\|\|^2 \leq \tfrac{1}{1-\beta^{n+1}(\alpha)}\langle \mathbf{T}_2(\alpha, n)\mathbf{u}^{(k+1)} - \mathbf{q}_2, \mathbf{u}^{(k+1)} - \mathbf{u}^*\rangle$$
$$\leq \tfrac{1}{1-\beta^{n+1}(\alpha)}\left(\varepsilon(u^{(k+1)})\right)^{\frac{1}{2}} \cdot \|\|\mathbf{u}^{(k+1)} - \mathbf{u}^*\|\|,$$

and the estimation (199) then follows. □

Let $\nu_1(\alpha)$ and $\nu_2(\alpha, n)$ denote the upper bounds for the rate of residual decreasing of the minimal residual solver applied respectively to equation (177) and (191). It follows from Theorem 3.6.1 and Theorem 3.6.2 that: for $\alpha > 0$ and $n \geq 0$,

$$\nu_1(\alpha) - \nu_2(\alpha, n) = \frac{(1 - \beta^{n+1}(\alpha)) + \beta(\alpha)(1 - \beta^n(\alpha))}{4} > 0. \tag{200}$$

It follows $\nu_1(\alpha) > \nu_2(\alpha, n)$ and the preconditioned minimal residual solver is theoretically faster than the minimal residual solver. Moreover, $\nu_2(\alpha, n) > \nu_2(\alpha, m)$ for $m > n$. Thus the convergence is theoretically more and more faster with increasing n. The Figure 35 plots estimates of upper bounds for the rate of residual decreasing as function of $\beta(\alpha)$ of the minimal residual method and its preconditioned versions for several values of n.

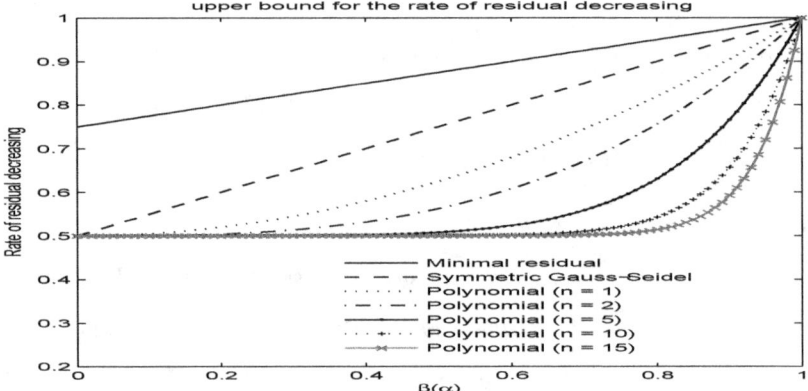

Figure 35: Comparison of upper bounds for the rate of residual decreasing of methods

Remark 3.6.1. *Since focus is on the solution of equation (168), for the computational purpose, we need only to solve the second sub-equation of problem (191) which reads:*

$$(I - M^{n+1}(\alpha))v = q_2, \tag{201}$$

where q_2 is the second component of vector $\mathbf{q_2}$. The solution u^* of (168) is computed from the the solution v^* of (201) as follows:

$$u^* = (\alpha I + A)^{-1}v^*. \tag{202}$$

Proceeding similarly as in the proof of Theorem 3.6.2 and Theorem 3.6.2, we have the following convergence results of the minimal residual method applied to (201):

$$\|q_2 - (I - M^{n+1}(\alpha))v^{(k+1)}\|^2 \leq \frac{1 + \beta^{n+1}(\alpha)}{2}\|q_2 - (I - M^{n+1}(\alpha))v^{(k)}\|^2, \tag{203}$$

$$\|v^{(k+1)} - v^*\| \leq \frac{1}{1 - \beta^{n+1}(\alpha)}\|q_2 - (I - M^{n+1}(\alpha))v^{(k+1)}\|. \tag{204}$$

3.6.4 Numerical Results for the 2-D Neutron Transport Equation

We apply the preceding minimal residual algorithm for solving neutron transport equation in 2-D cartesian geometry, describes in section 3.5.3.

The discretization is carried out by a DSN scheme (see [13, 34]) consisting of using a finite set of L discrete angular directions $\Omega_L = \{\Omega_i = (\mu_i, \eta_i)\}_{i=1}^{i=L} \subset B$, which have nonzero components and are symmetric about the origin for the angular approximation and, a difference method based on control volume approach and cell averaging for the spatial approximation. The numerical grid is defined by:

$$R_h = \{(x_i, y_j), 0 \le i \le N, 0 \le i \le M\}, \tag{205}$$

where $x_0 = 0$, $x_i = x_{i-1} + (\Delta x)_i$, $x_N = 1$, $y_0 = 0$, $y_j = y_{j-1} + (\Delta y)_j$, $y_M = 1$ and $h = \max_{ij} ((\Delta x)_i, (\Delta y)_j)$. The cell center grid points are defined as: $x_{i+\frac{1}{2}} = \dfrac{x_{i+1} - x_i}{2}$, $y_{j+\frac{1}{2}} = \dfrac{y_{j+1} - y_j}{2}$, $(\Delta x)_{i+\frac{1}{2}} = x_{i+1} - x_i$ and $(\Delta y)_{j+\frac{1}{2}} = y_{j+1} - y_j$.

Therefore, equations (194) and (195) can be solved using respectively a direct sweeping algorithm (see [13, 34]) and a conjugate gradient method in the anisotropic case (see [13]).

For the numerical results, we took particular data for which an exact solution of problem (159) is known in each case. For the iterative methods tested here, the iterations are stopped when the relative error $\|U - U_{exate}\|_2 / \|U_{exate}\|_2$ is less than a prescribed $\epsilon > 0$.

For $x = (x_1, x_2) \in R$ and $\Omega = (\mu, \eta) \in B$, we set $\sigma(x) = \sigma$, $\kappa(x, \Omega, \Omega') = \frac{\sigma c}{\pi}$ and

$$q(x, \mu) = \begin{cases} \mu x_2 + \eta x_1 + \sigma x_1 x_2 - \frac{\sigma c}{4}, & \mu > 0, \eta > 0; \\ -\mu x_2 + \eta(1 - x_1) + \sigma(1 - x_1)x_2 - \frac{\sigma c}{4}, & \mu < 0, \eta > 0; \\ -\mu(1 - x_2) - \eta(1 - x_1) + \sigma(1 - x_1)(1 - x_2) - \frac{\sigma c}{4}, & \mu < 0, \eta < 0; \\ \mu(1 - x_2) - \eta x_1 + \sigma x_1(1 - x_2) - \frac{\sigma c}{4}, & \mu > 0, \eta < 0. \end{cases}$$

The exact solution of this test problem is given by:

$$\psi(x, \mu) = \begin{cases} x_1 x_2, & \mu > 0, \eta > 0; \quad (1 - x_1)x_2, \mu < 0, \eta > 0; \\ (1 - x_1)(1 - x_2), & \mu < 0, \eta < 0; \quad x_1(1 - x_2), \mu > 0, \eta < 0. \end{cases}$$

In this problem, c is the scattering ratio and σ is the optical coefficient. The quantities $\sigma_s = \sigma c$ and $\sigma_a = \sigma(1 - c)$ are respectively the scattering and absorption cross sections of neutrons in R.

For the numerical test, we take $\Delta x = \Delta y = \frac{1}{10}$ and L=100. We study the behavior of the preconditioned minimal residual methods with respect to parameter σ, c and α. For the example problem, the theoretical optimal parameter which minimizes the bound $\beta(\alpha)$ is $\alpha_t = \sigma_a = \sigma(1 - c)$. It is observed in [13] that, for fixed c and σ, numerical optimal value of α can be localized in the interval $[\sigma(1 - c), \sigma(1 - c/2)]$. The value

$$\alpha* = \begin{cases} \sigma(1 - 19c/32), & 0 < c < 0.9 \\ \sigma(1 - 23c/32), & 0.9 \le c < 0.97 \\ 1, & 0.97 \le c \le 1 \end{cases}$$

yielded good convergence results for the SAS iteration applied to the example problem as compare to the standard source iteration method, a spatial multigrid algrithm and some Krylov subspace methods such as GMRES and BiCGStab iterative algorithms [13]. It is observed in [16] that for some values of σ_a and c ($\sigma_a \geq 4$ and $c > 0.5$.), the convergence of the minimal residual seemed to be faster as compare to the SAS method using α_t. This results also holds for large value of σ when using the SAS method with $\alpha*$. The Gauss-Seidel preconditioned version of the minimal residual algrithm gave excellent results compare to SAS and its Successive overrelaxation acceleration (see [16]).

We presented comparative numerical results (number of iterations and CPU time in s) of the previous minimal residual algorithm with: symmetric Gauss-Seidel preconditioning (SGS-Minres), the polynomial preconditioning (PMinres[n]) with n denoting the order of the truncated Neumann series and the SAS iterations using $\alpha = \alpha_*$ and $\alpha = \alpha_t + c$. There is two sets of tests: one at fixed σ and another at fixed c. As shown by Figure 36 to Figure 42, all the methods converge. At fixed $\sigma = 50$ and $\sigma = 100$, we compare the c−dependence of the iterative methods used here. The Figure 36 and Figure 37 plot the number of iterations and CPU time of methods as function of c respectively for $c \in [0.1, 0.99]$ and $c \in [0.98, 0.99999]$ with $\sigma = 50$, using $\alpha = \alpha_*$. We observe That the PMinres[n] iterations ($n = 1, 2, 10$] are faster than the SGS-Minres which is faster than the SAS, particularly for values of c closed to 1 (Figure 37). As we can see from Figure 38 and Figure 39, these observations remain true when performing the same tests for $\sigma = 100$. Next, we compare the σ−dependence of iterative methods. The Figure 40 and Figure 41 plot the number of iteration of the methods and the CPU time as function of the total cross section ($\sigma \in [1, 100]$) at fixed $c = 0.5$ and $c = 0.99$ respectively. We can see that the PMinres[n] method is still more efficient than the SGS-Minres which is faster than SAS method. The same observations hold for large values of σ at $c = 0.99$ (Figure 42) and for the critical case where $c = 1$ (Table 5). We set $\alpha = \sigma(1 - c) + c$. Table 6 and Table 7 present comparative numerical results of methods for $1 \leq \sigma \leq 1000$ at fixed $c = 0.98$ and for $0.98 \leq c \leq 1$ at fixed $\sigma = 5$ respectively. The SAS method remain slower than the preconditioned minimal residual methods. We also remark that for $\alpha = \alpha^*$, PMinres[n] is more and more efficient with increasing n. This confirms the theoretical convergence results obtained.

We now consider another set of tests where the mesh size decreases : $\Delta x = \Delta y = h$, with $h \in \{\frac{1}{4}, \frac{1}{8}, \frac{1}{16}, \frac{1}{32}, \frac{1}{64}, \frac{1}{128}\}$. At fixed $\sigma = 100$, we test the behavior of the methods as mesh size decreases, for $c = 0.5$ and $c = 0.99$. The Table 8 presents the number of iterations of each method for $c = 0.5$. It can be observed that for each method, the number of iterations is roughly constant, as the mesh size decreases. For $c = 0.99$, we set $\alpha = \sigma(1 - 23c/32)$. The Table 9 presents the number of iteration of each methods. The tested methods converge for a mesh size less than $\frac{1}{64}$. For $h = \frac{1}{128}$, the convergence of SGS-Minres and PMinres[1] is denoted. For the others methods, we observe the divergence of the SAS method at the second iteration and the stagnation of the residual for Pminres[2] and Pminres[10]. This drawback is essentially due to the fact that the discretization method applied to the first subproblem of the SAS iteration generates negative flux for the fixed values of σ, c and α. This drawback is overcome by setting $\alpha = \sigma$. The Table 10 gives the number of iterations of methods tested here. It can be observed that for each methods, the number of iterations is roughly constant for $h \geq \frac{1}{16}$ and the preconditioned Minres methods accelerate the SAS iterations. The convergence behavior (relative error as function of iteration) of the SAS, SGS-Minres, PMinres[n] ($n = 1, 2, 10$) is plotted in Figure 43 for $h = \frac{1}{128}$, $\sigma = 100$ and $c = 1$ with $\alpha = \sigma$. The efficiency of the preconditioned minimal residual methods can be observed.

Additionally, we present comparative convergence behavior of preconditioned Minres methods and a spatial multigrid method using the bi-conjugate gradient stabilized method as smoothing method MG(n_1,n_2), n_1 and n_2 denoting the number of pre-smoothing and post-smoothing steps respectively. The iteration are stopped when the relative residual error $\|B - GU\|/\|B\|$ is less than $1E - 05$, where G and B denote respectively, the matrix and the right hand side of the discrete system . The convergence history of multigrid and SGS-Minres method for $\sigma = 100$ and $c = 0.5$ with $h = \frac{1}{16}$ is plotted in Figure 44. It can be seen that the MG(1,1) method diverges and MG(20,20) method converges but is less efficient than the SGS-Minres. We set $c = 0.99$. It is observed in Figure 45 that, for the mesh size $h = \frac{1}{16}$, the MG(1,1) method diverges and the preconditioned Minres methods are efficient compare to MG(20,20) method. The Figure 46 plots the convergence behavior of the methods tested here for the mesh size $h = \frac{1}{32}$, at fixed $c = 0.99$ and $\sigma = 100$. The divergence of MG(1,1) and MG(20,20) methods can be observed. We denote efficiency of the preconditioned Minres methods compare to the spatial multigrid methods considered.

We now set $\Delta x = \Delta y = 1/15$. The approximate scalar flux is computed and compared to the exact scalar flux which is $\Phi(x, y) = \pi/4$. The Figure 47 plots the approximation error of the scalar flux during iterations using PMinres[3] and SAS method when $\sigma = 50$

121

and $c = 1$. We notice that PMinres[3] iteration gives excellent results compare to SAS iteration.

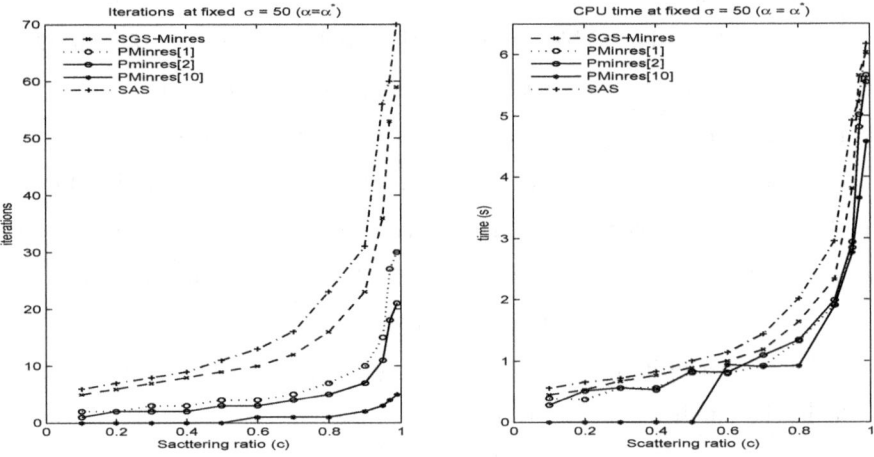

Figure 36: Comparison of methods at fixed $\sigma = 50$, for $c \in [0.1, 0.99]$ ($\epsilon = 1E - 08$): (left) number of iteration;(right) CPU time in s.

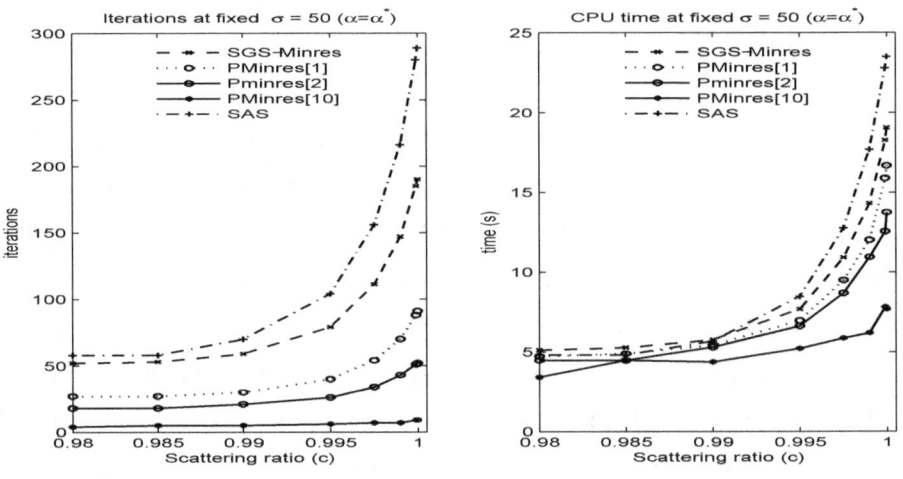

Figure 37: Comparison of methods at fixed $\sigma = 50$, for $c \in [0.98, 0.99999]$ ($\epsilon = 1E - 08$): (left) number of iteration;(right) CPU time in s.

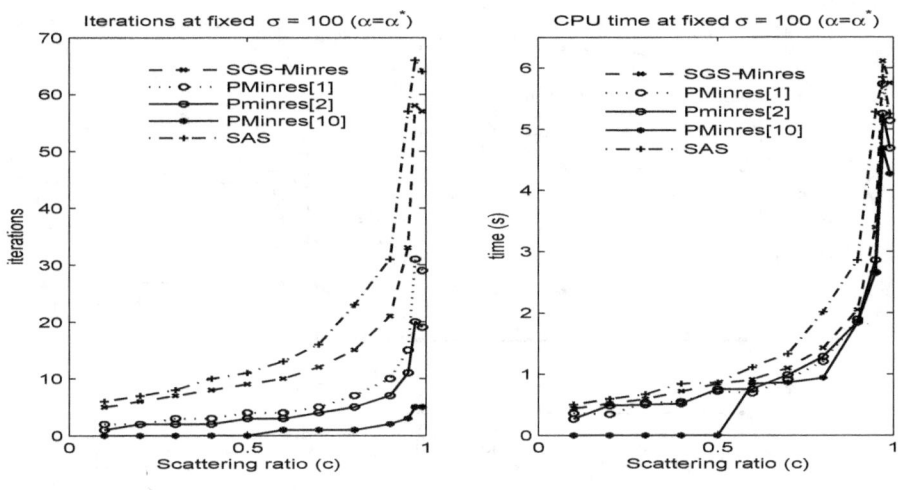

Figure 38: Comparison of methods at fixed $\sigma = 100$, for $c \in [0.1, 0.99]$ ($\epsilon = 1E - 08$): (left) number of iteration;(right) CPU time in s.

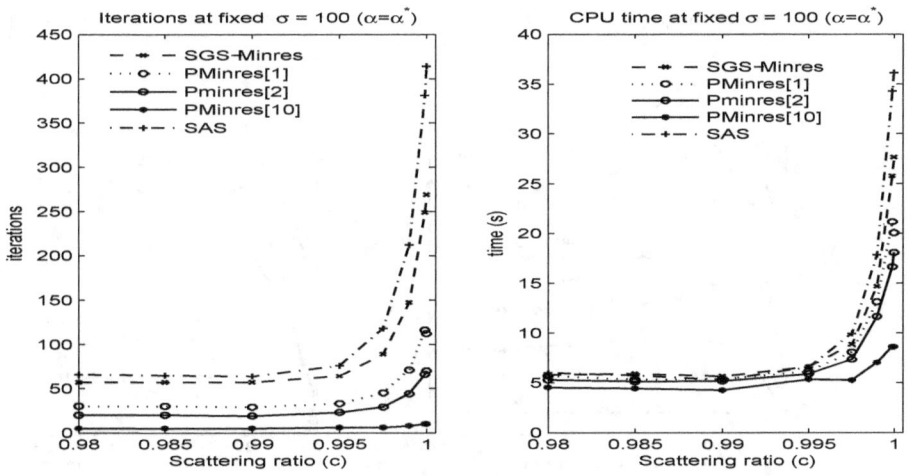

Figure 39: Comparison of methods at fixed $\sigma = 100$, for $c \in [0.98, 0.99999]$ ($\epsilon = 1E-08$): (left) number of iteration;(right) CPU time in s.

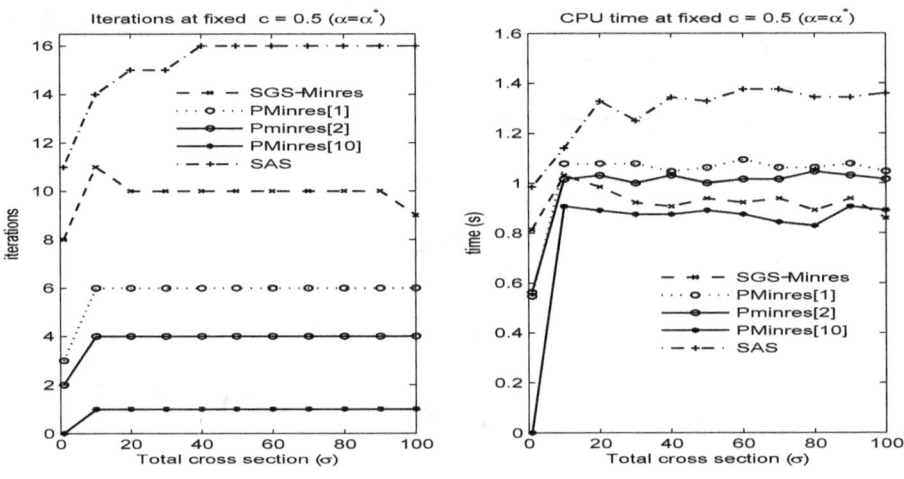

Figure 40: Comparison of methods at fixed $c = 0.5$, for $\sigma \in [1, 100]$ ($\epsilon = 1E - 08$): (left) number of iteration;(right) CPU time in s.

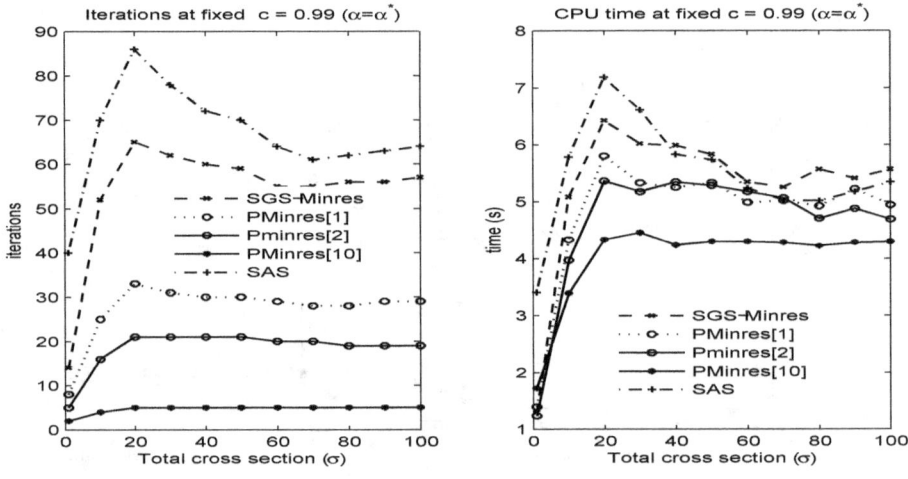

Figure 41: Comparison of methods at fixed $c = 0.99$, for $\sigma \in [1, 100]$ ($\epsilon = 1E - 08$): (left) number of iteration;(right) CPU time in s.

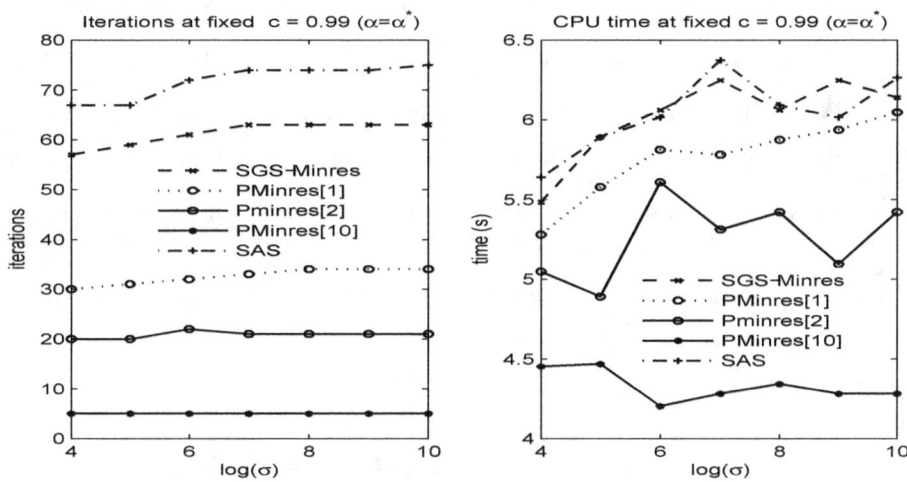

Figure 42: Comparison of methods at fixed $c = 0.99$, for large σ ($\epsilon = 1E - 08$): (left) number of iteration;(right) CPU time in s.

σ	1	10	50	100	500	1000
SAS	42(3.97)	86(7.16)	290(24.45)	418(34.95)	648(55.11)	696(58.03)
SGS-Minres	14(1.30)	61(6.27)	191(19.97)	271(28.22)	416(40.59)	446(44.41)
PMinres[1]	8(1.41)	28(4.95)	91(15.92)	112(20.06)	193(34.34)	204(35.42)
PMinres[2]	5(1.22)	19(4.80)	53(13.23)	71(17.69)	105(26.37)	112(28.86)
PMinres[10]	2(1.78)	4(3.37)	9(7.75)	10(8.45)	12(10.37)	12(10.16)

Table 5: Iterations Number and CPU time in s (in bracket) for $c = 1$ ($\alpha = 1$, $\epsilon = 1E - 08$)

σ	1	10	50	100	500	1000
SAS	38(3.53)	46(4.11)	43(3.70)	59(4.85)	185(15.97)	276(23.69)
SGS-Minres	14(1.44)	39(4.06)	42(4.17)	40(3.92)	24(2.28)	19(1.89)
PMinres[1]	8(1.56)	18(3.44)	21(3.61)	27(5.01)	70(12.06)	95(16.94)
PMinres[2]	5(1.36)	12(3.44)	14(3.48)	18(4.48)	23(5.87)	24(6.23)
PMinres[10]	2(1.98)	3(2.90)	3(2.59)	5(4.31)	15(12.80)	18(16.56)

Table 6: Iterations Number and CPU time in s (in bracket) for $c = 0.98$ ($\alpha = \sigma(1-c)+c$, $\epsilon = 1E-08$)

c	0.98	0.99	0.995	0.9975	0.999	0.9999	1
SAS	39(3.72)	42(4.62)	49(5.61)	56(6.26)	60(6.56)	62(6.92)	63(6.80)
SGS-Minres	31(3.31)	33(4.20)	35(4.70)	36(4.70)	36(4.81)	36(4.73)	37(4.94)
PMinres[1]	14(2.81)	15(3.62)	17(4.20)	18(4.33)	19(4.39)	20(4.80)	20(4.87)
PMinres[2]	9(2.55)	10(3.31)	11(3.69)	11(3.72)	11(3.67)	11(3.64)	11(3.69)
PMinres[10]	3(2.83)	3(3.58)	3(3.39)	3(3.44)	3(3.47)	3(3.55)	3(3.43)

Table 7: Iterations Number and CPU time in s (in bracket) for $\sigma = 5$ ($\alpha = \sigma(1-c)+c$, $\epsilon = 1E-08$)

h	$\frac{1}{4}$	$\frac{1}{8}$	$\frac{1}{16}$	$\frac{1}{32}$	$\frac{1}{64}$	$\frac{1}{128}$
SAS	8	8	8	8	8	8
SGS-Minres	6	6	7	7	7	7
PMinres[1]	3	3	3	3	3	3
PMinres[2]	2	2	2	2	2	2
PMinres[10]	0	0	0	0	0	0

Table 8: Iterations Number for $\sigma = 100$, $c = 0.5$ ($\alpha = \alpha^*$, $\epsilon = 1E-06$)

h	$\frac{1}{4}$	$\frac{1}{8}$	$\frac{1}{16}$	$\frac{1}{32}$	$\frac{1}{64}$	$\frac{1}{128}$
SAS	182	184	186	186	186	-
SGS-Minres	49	78	91	99	128	205
PMinres[1]	26	32	35	37	46	88
PMinres[2]	18	19	20	21	24	-
PMinres[10]	6	6	6	7	7	-

Table 9: Iterations Number for $\sigma = 100$, $c = 0.99$ ($\alpha = \sigma(1 - 23c/32)$, $\epsilon = 1E - 06$)

h	$\frac{1}{4}$	$\frac{1}{8}$	$\frac{1}{16}$	$\frac{1}{32}$	$\frac{1}{64}$	$\frac{1}{128}$
SAS	633	637	640	641	642	643
SGS-Minres	20	34	46	51	49	53
PMinres[1]	7	21	32	32	33	33
PMinres[2]	7	18	26	26	26	26
PMinres[10]	7	13	14	14	13	13

Table 10: Iterations Number for $\sigma = 100$, $c = 0.99$ ($\alpha = \sigma$, $\epsilon = 1E - 06$)

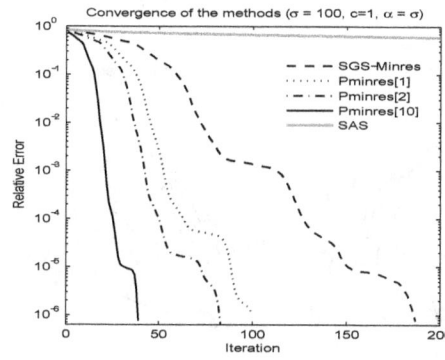

Figure 43: Convergence behavior of Preconditioned Minres and SAS Methods for $h = \frac{1}{128}$, $\sigma = 100$ and $c = 1$ with $\alpha = \sigma$.

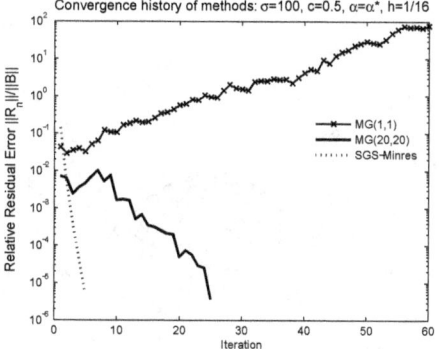

Figure 44: Convergence behavior of Multigrid and SGS-Minres Methods for $h = \frac{1}{16}$, $\sigma = 100$ and $c = 0.5$ with $\alpha = \alpha*$.

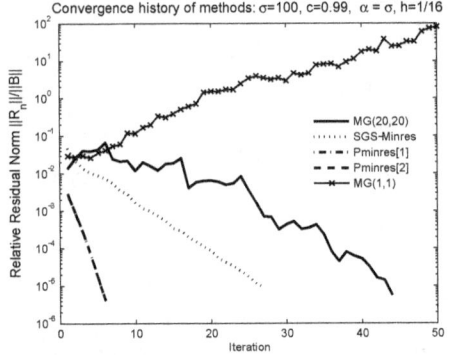

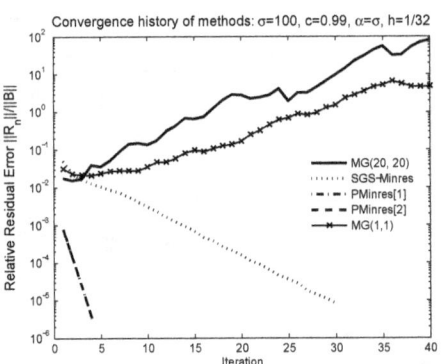

Figure 45: Convergence behavior of Multigrid and Preconditioned Minres Methods for $h = \frac{1}{16}$, $\sigma = 100$ and $c = 0.99$ with $\alpha = \sigma$.

Figure 46: Convergence behavior of Multigrid and Preconditioned Minres Methods for $h = \frac{1}{32}$, $\sigma = 100$ and $c = 0.99$ with $\alpha = \sigma$.

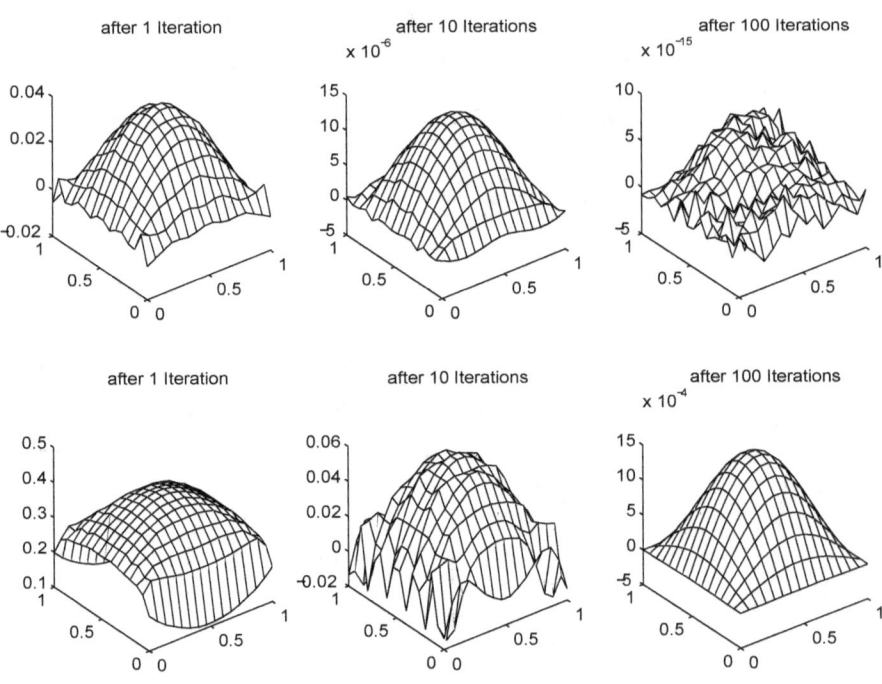

Figure 47: Error in scalar flux approximation using PMinres[3] and SAS methods for $\sigma = 50$ and $c = 1$ ($\Delta x = \Delta y = 1/15$): (top) PMinres[3] iteration; (bottom) SAS iteration.

128

3.7 Summary

Here we summarize the results developed in this chapter.

The two-step iterative method (PAS) for solving linear operator equations with operators admitting positive definite and m-accretive splitting in a Hilbert space H converges. The analysis of a SOR acceleration of this method yields convergence results similar to those obtained in finite dimensional linear systems with coefficient matrices possessing Property A. Numerical results have shown the convergence of the PAS iteration and have demonstrated the effectiveness of the SOR acceleration of the PAS method in solving an integro-differential transport equation in 1-D spatial geometry.

The iterative methods based on a Self-adjoint and m-Accretive splitting of the transport operator presented, converge unconditionally to the solution of the transport equation. The previous numerical results show the feasibility and the effectiveness of the SAS iteration. A bound for the contraction factor of the SAS iteration have been derived. It appears that the SAS iteration is efficient compare to the standard Source Iteration. The method converges for critical cases (c close to 1 and/or large σ) and is easy to implemented in dimension one as well as in dimension two. Furthermore the numerical results have demonstrated the effectiveness of the SOR acceleration of the SAS method for solving the neutron transport problem in 2-D geometry and in 1-D spherical geometry. It is important to mention that the theoretical proof of the convergence of the method is independent of the discretization.

The convergence of a minimal residual method for the solution of a class of linear operator equation,with positive definite operator admitting the Self-adjoint and m-Accretive Splitting in a Hilbert space H have been analyzed . A bound for the rate of residual decreasing have been derived. Numerical results illustrate the convergence of the method for the solution of a 2-D neutron transport problem. Moreover, the Gauss-Seidel preconditioning of this minimal residual solver gives excellent results than the SAS method and its SOR acceleration.

We have presented a splitting iterative method for the solutions of a class of linear operator equation, with positive definite operator admitting the Self-adjoint and m-Accretive Splitting in a Hilbert space H. Theoretical analysis of the symmetric Gauss-Seidel and polynomial preconditioning of the previous minimal residual method shown that these method convergence unconditionally to the solution of the equation. Theoretical proof of the convergence of methods is independent of the discretization. Numerical results illustrate feasibility and the effectiveness of these methods for solving a 2-D neutron transport problem. The methods converge for critical cases (c close to 1 and/or large σ). Moreover, the above preconditioned Minres methods give excellent results than the SAS iteration method.

General Conclusion

Conclusion

In this thesis we have derived new classes of (ADI)-like iterative methods based on positive definite and m-accretive splitting (PAS) for the treatment of the steady state neutron transport equation in slab geometry, bounded convex domain of $\mathbb{R}^n$ ($n = 2, 3$) and in 1-D spherical geometry. Theoretical analysis have shown that the PAS iteration method converges unconditionally and its SOR acceleration yields convergence results similar to those obtained in presence of finite dimensional systems with matrices possessing the *Young property A*. For the particular case where the positive definite part of the linear equation operator is self-adjoint, we have derived an upper bound for the contraction factor of the iterative method, which depends solely on the spectrum of the self-adjoint part. The incomplete version of SAS iteration has been analyzed. The convergence analysis of an infinite dimensional adaptation of minimal residual and preconditioned minimal residual algorithms using Gauss-Seidel, symmetric Gauss-Seidel and polynomial preconditioning applied to solve the 2 by 2 matrix operator equation resulting from the self-adjoint and m-accretive splitting has been provided. Upper bounds for The rates of residual decreasing of these minimal residual methods which depend solely on the spectrum of the self-adjoint part have been derived. The convergence of theses solvers was illustrated numerically on sample neutron transport problems in slab geometry, in 2-D cartesian geometry and in 1-D spherical geometry. Various test cases, including pure scattering and optically thick domains are considered.

Future Recommendations

The methods described in this dissertation may be further improved and developed by studying the following issues :

- The Fourier analysis of the convergence of the method.

- The analysis of the two parameters version which uses different parameters for the two half iterations in each iteration step.

- The determination of the numerical optimal parameter.

- The analysis of non-stationary version of the methods.

- The acceleration of the method by preconditioning, synthetic acceleration and multi-grid approaches.

Bibliography

[1] N. G. Abrashina-Zhadaeva and A. A. Egorov, *Multicomponent iterative methods solving stationary problems of mathematical physics* Mathematical Modelling and Analysis, **13** (2008), no. 3, pp. 313–326. (doi:10.3846/1392-6293.2008.13.313-326)

[2] R. T. Ackroyd, *Finite Element Methods for Particle Transport. Applications to reactor and radiation physics*, Research Studies in Particle and Nuclear Technology, **6**, Research Studies Press Ltd, Taunton, 1997.

[3] R. T. Ackroyd, *Foundation of Finite Element Applications to Neutron Transport*, Progress in Nuclear Energy, **29** (1995), No. 1, pp. 43-56.

[4] R. T. Ackroyd, J. K. Fletcher, A. J. H. Goddard, J. Issa, M. M. R. Williams and J. Wood, *Some Recent Development in Finite Element Method for Neutron Transport*, Advance in Nuclear Science and Technology, **19** (1987), pp. 381-483.

[5] M. L. Adams and E. W. Larsen, *Fast Iterative Methods for Discrete-Ordinates Particle Transport Calculation*, Prog. Nucl. Energy, **40** (2002), pp. 3–159.

[6] S. Akesbi and E. Maître, *Theoretical and Numerical Analysis of a Minimal Residual Solver for 2D Boltzmann equation*, Journal of Computation and Applied Mathematics, **150** (2003), no. 2, pp. 357-374.

[7] S. Akesbi and E. Maître, *Minimal Residual Method Applied to the Transport equation*, Journal of Numerical Algorithms, **26** (2001), pp. 235-249.

[8] G. Alpert, G. Beyklin, D. Gines and L. Vozovoi, *Adaptive solution of partial differential equations in multiwavelet bases*, J. Comput. Phys., **182** (2002), pp. 149-190.

[9] X. Antoine and M. Lemou, *Wavelet approximation of a collision operator in kinetic theory*, C. R. Acad. Sci. Paris, **Ser.I 337** (2003), pp. 353-358.

133

[10] A. B. Antonevich, J. Appell, V. A. Prokhorov, and P. P. Zabrejko, *Quasi-iteration methods of Chebyshev type for the approximate solution of operator equations*, REND. SEM. MAT. UNIV. PADOVA, **93** (1995), pp. 127–141.

[11] T. M. Austin, *Advance on a Scaled Least-Squares Method for the 3-D Linear Boltzmann Equation*, PhD thesis, University of Colorado, USA, 2001.

[12] O. Awono and J. Tagoudjeu. *Iterative Methods for a Class of Linear Operator Equations* Int. J. Contemp. Math. Sci., **4** (2009), no. 12, pp. 549–564.

[13] Awono Onana and J. Tagoudjeu, *A Splitting Iterative Method for Solving the Neutron Transport Equation*, Mathematical Modelling and Analysis, **14** (2009), no. 3, pp. 271–289. (doi:10.3846/1392-6292.2009.14.271-289)

[14] O. Awono and J. Tagoudjeu, *A Self-Adjoint and m-Accretive Splitting Iterative Method for Solving the Neutron Transport Equation in 1-D Sphérical Geometry*, Proceeding of the 9th African Conference on Research in Computer Science and Applied Mathematics, Morocco, (2008), pp. 331-338.

[15] O. Awono and J. Tagoudjeu, *A SOR acceleration of Self-Adjoint and m-Accretive Splitting Iterative Solver for 2-D Neutron Transport Equation*, Proceeding of the 9th International Conference JANO'9 , Mohammedia-Morocco, (2008), pp. 318-321.

[16] O. Awono and J. Tagoudjeu. *A Minimal Residual Solver for the Neutron Transport Equation*, Int. J. Contemp. Math. Sci., **4** (2009), no. 34, pp. 1671–1684.

[17] O. Awono and J. Tagoudjeu. *A Preconditioned Minimal Residual Solver for a Class of Linear Operator Equations*, to appear in: Computational Methods in Applied Mathematics.

[18] O. Awono, S. Yunkap Kwankam and J. Mvogo Ngono. *Solution of the two-dimensional neutron transport equation by a hierarchical finite element method*, Internat. J. Numer. Methods Heat Fluid Flow, **3** (1993), pp. 35–47,

[19] M. Azadzadeh, *Convergence of a Discontinuous Galerkin Scheme for the Neutron Transport*, Transport Theory Statist. Phys., **30** (2001), No.4-6, pp. 357-383.

[20] Z.-Z. Bai, G.H. Golub and M.K. NG, *Hermitian and Skew-Hermitian Splitting Methods for Non-Hermitian Positive Definite Linear Systems*, SIAM Journal on Matrix Analysis and Applications, **24** (2003), pp. 603-626.

[21] Z.-Z. Bai, G.H. Golub and M.K. NG, *On Successive Overrelaxation of Hermitian and Skew-Hermitian Splitting Methods for Non-Hermitian Positive Definite Linear Systems*, Technical Report SCCM02-06, Stanford University, (2002)

[22] R. Barrrett, M. Berry, T. F. Chan, J. Demmel, J. Donato, J. Dongarra, V. Eijkhout, R. Pozo, C. Romine and H. Van der Vorst. *Templates for the Solution of Linear Systems: Building Blocks for Iterative Methods.* SIAM, Philadelphia, 1994.

[23] L. Bourhrara, *New Variational Formulations for the Neutron Transport Equation*, Transport Theory and Statistical Physics, **33**, (2004), No. 2, pp. 93–124

[24] V. Brattka and R. Dillhage, *Computability of the Spectrum of Self-Adjoint Operators*, Journal of Universal Computer Science, **11**, (2005), pp. 1884–1900

[25] H. Brezis, *Analyse Fonctionnelle. Theorie et Application*, 2ème tirage, Masson, Paris, 1987.

[26] A.G. Buchan, C.C. Paina, M.D. Eaton, R.P. Smedley-Stevensonb and A.J.H. Goddard *Linear and Quadratic Octahedral Wavelets on the Sphere for Angular Discretisations of the Boltzmann Transport Equation*, Annals of Nuclear Energy, **32** (2005), pp. 1224-1273.

[27] A. Buchan, *Adaptive Spherical Wavelets for the Angular Discretisation of the Boltzmann Transport Equation*, Ph.D Dissertation, Imperial College of Science, Technology and Medecine, University of London, 2006.

[28] L. Cao, H. Wu and Y. Zheng *Solution of neutron transport equation using DaubechiesŠ wavelet expansion in the angular discretization*, Nucl. Eng. Des. (2008), doi:10.1016/j.nucengdes.2008.03.003.

[29] J. Cartier, *Résolution de l'Équation du Transport par une Méthode d'éléments Finis Mixtes-Hybrides et Approximation de la Diffusion par de Problèmes de Transport*, Thèse de Doctorat de l'Université d'Orleans, France, 2006.

[30] B. Chang and B. Lee. *A multigrid algorithm for solving the multi-group anisotropic scattering boltzmann equation using first-order system least-squares methodology*, Electronic Transactions on Numerical Analysis, **15** (2003), pp. 132–151.

[31] R. Čiegis, O. Iliev and Z. Lakdawala, *On Parallel Numerical Algorithms for Simulating Industrial Filtration Problems*, Computational Methods in Applied Mathematics, **2**, (2007), No. 2, pp. 118–134

[32] Raim. Čiegis, Rem. Čiegis and M. Meilūnas, G. Jankevičiūtė, V. Starikovičius, *Parallel numerical algorithm for optimization of electrical cables*, Mathematical Modelling and Analysis, **13**, (2008), No. 4, pp. 471–482

[33] Adrian C. Constantinescu, *Analysis of Projective-Iterative Methods for Solving Multidimensional Transport Problems* MS Thesis, Graduate Faculty, North Carolina State University, (2008)

[34] R. Dautray and J.-L. Lions, *Analyse Mathématique et Calcul Numérique pour les Sciences et les Techniques*, Tome 3, Masson, Paris, 1985.

[35] M. D. DeHart, *A Discrete Ordinates Approximation to the Neutron Transport Equation Applied to Generalized Geometries*: Ph.D Dissertation, Office of Graduate Study, Texas A&M University, 1992.

[36] J. R. Dishaw, *Time Dependent Discrete Ordinates Neutron Transport using Distribution Iteration in XYZ Geometry*, Ph.D Dissertation, Air Force Institute of Technology, Air University, 2007.

[37] J.J. Duderstadt, E.E. Lewis and C. Bardos, *Neutron Transport Equation*, Eyrolles, Paris, 1983.

[38] B. B. Ganapol, *Analytical Benchmark for Nuclear Engineeering Application. Case Study in Neutron Transport Theory*, Report NEA No. 6292, NEA-OECD, 2008.

[39] M. G. Gasparo, A. Papini, and A. Pasquali, *Some properties of GMRES in Hilbert spaces*, Numer. Funct. Anal. Optim., **29** (2008), No. 11-12, pp. 1276–1285.

[40] C. J. Gesh, *Finite Element Methods for Second order Forms of the Transport Equation*: Ph.D Dissertation, Texas A&M University, 1999.

[41] P. Guérin, *Méthodes de Décomposition de Domaine pour la Formulation Mixte Duale du Problème Critique de la Diffusion des Neutrons*, Thèse de Doctorat, Université Paris VI, France, Décembre 2007.

[42] A. Kadem, *Spectral Methods for the Transport Equation*, Ph.D Thesis, Departement of Mathematics, Univerty of Setif, 2006.

[43] L. Knizhnerman, *On GMRES-Equivalent Bounded Operator*, SIAM J. Matrix Anal. Appl., **23** (2000), no. 1, pp. 195–212.

[44] K. Kobayashi, N. Sugimura and Y. Nagaya, *3-D Radiation Transport Benchmark Problems and Results for Simple Geometries with Void Regions*, Report, Nuclear Science Committee, NEA-OECD, 2000.

[45] M.A. Krasnosel'skii, G. M. Vainikko, P. P. Zabreiko, Ya. B. Rutitskii and V. Ya. Stetsenko, *Approximate Solution of Operator Problem*, Wolters-Noordhoff Publishing, Groningen, 1972.

[46] B. D. Lansrud, *A Spatial Multigrid Iterative Method for Two-Dimensional Discrete-Ordinates Transport Problems*, Ph.D Dissertation, Texas A&M University, 2005.

[47] E. W. Larsen, *An Overview of Neutron Transport Problems and Simulation Techniques*: in Computational Methods in Transport, pp. 513–534, Lect. Notes Comput. Sci. Eng., **48**, Springer Berlin, 2006.

[48] E. W. Larsen, *The Description of Particles Transport Problems with Helical Symmetry*, Nuclear Science and Engineering, **156** (2007), pp. 68–73.

[49] E. W. Larsen and J. E. Morel, *Advances in Discrete-Ordinates Methodology*: in Nuclear Computational Science: A century of Review, Eds. Y. Amzi and E. Sartori, Springer Berlin, 2006.

[50] P. Lascaux and R. Théodor, *Analyse Numérique Matricielle Appliquée à l'Art de l'Ingénieur*, Volume 2, Masson, Paris, 1987.

[51] J. Leppänen, *Development of a New Monte Carlo Reactor Physics Code*, PhD thesis, Helsinki University of Technology, Finland, June 2007.

[52] Lewis, E. E. (1981) *Finite Element Approximation to the Event-parity Transport Equation*, Nuclear Science and Technology, **13** (1981), pp. 155-225.

[53] G. Longoni, *Advance Quadrature Sets, Acceleration and Preconditioning Thechniques for the Discrete Ordinates Method in Parallel Computing Environements*: Ph.D Dissertation, Graduate School, University of Florida, 2006.

[54] P. Malits, *Certain approximate methods for solving linear operator equations*, Appl. Math. Lett., **20** (2007), no. 3, pp. 306–311.

[55] T. A. Manteuffel, S. McCormick, J. E. Morel, S. Oliveira and G. Yang. *A Parallel Version of A multigrid algorithm for isotropic transport Equations*. SIAM Journal on Scientific Computing, **15**, (1994), no. 2, pp. 474–493.

[56] T. A. Manteuffel, S. McCormick, J. E. Morel and G. Yang. *A fast multigrid algorithm for isotropic transport problems ii. with absorption*. SIAM Journal on Scientific Computing, **17**, (1996), pp. 1449–1474.

[57] T. A. Manteuffel and K. Ressel. *A Systematic Solution Approach for Neutron Transport Problems in Diffusive Regimes,* in Seventh Copper Mountain Conference on Multigrid Methods , N. D. Melson, T. A. Manteuffel, S. McCormick, and C. C. Douglas, eds., NASA Hampton, VA, (1996), pp. 519–534.

[58] T. A. Manteuffel and K. Ressel. *Least-squares finite element solution for the neutronic transport equation in diffusive regimes*. SIAM J. Numer. Anal., **35** (1998), pp. 806–835.

[59] G. Marchuk. *Decomposition Methods . Nauka, Moscou, 1988. (In Russian)*

[60] G. I. Marchouk and V. Agochkov, (1981), *Kinetic Equations and Variational principles*, SIAM J. Numer. Anal., **18** (1981), No. 2

[61] G. Marchuk and V. Agochkov, *Introduction aux Méthodes des Eléments Finis*, Mir, Moscou, 1985.

[62] A. E. Maslowski Olivares, *A New Iterative Approach to Solving the Neutron Transport Equation*, Ph.D Dissertation, Office of Graduate Study, Texas A&M University, 2008.

[63] L. Mei, *Multigrid algorithms for solving multigroup neutron transport,* Applied Mathematics and Computation **179** (2006), pp. 473–483.

[64] J. E. Morel and J. M. McGhee, *A self-adjoint angular flux equation,* Nucl. Sci. Eng. **179** (1999), pp. 312Ũ325.

[65] O. Nevanlinna, *Convergence of Krylov methods for sums of two operators,* BIT, **36** (1996), no.4, pp. 775–785.

[66] S. Oliveira and Y. Deng. *Preconditionrd krylov subspace methods for transport equations,* Prog. Nucl. Energy, **33** (1998), No. 1/2, pp. 155–174.

[67] B. W. Patton and J. P. Holloway. *Preconditioned gmres to the numerical solution of the neutron transport equation,* Ann. Nucl. Energy, **29**, (2002).

[68] A. G. Ramm, *Iterative solution of linear equations with unbounded operator,* J. Math. Anal. Appl., **330** (2007), pp. 1338-1346.

[69] S. Richling, E. Meinköhn, N. Kryzhevoi and G. Kanschat, *Radiative transfer with finite element I. Basic method and tests,* Astronomy & Aastrophysics, **380** (2001), pp. 776–788. doi: 10.1051/0004-6361:20011411

[70] R. Sanchez and N.J. McCormick, *A Review of Neutron Transport Approximations,* Nucl. Sci. Eng. **80** (1982), pp. 481–535.

[71] R. Sanchez and N.J. McCormick, *A Discrete ordinates solutions for highly forward-peaked scattering,* Nucl. Sci. Eng. **147** (2004), pp. 249–274.

[72] F. Schäpfer, A. K. Louis, and T. Schuster, *Nonlinear iterative methods for linear ill-posed problems in banach spaces,* Inverse Problems, **22** (2006), pp. 311Ũ329.

[73] M. Seaïd, *A Note on Numerical Methods for Two-Dimensional Neutron Transport Equation,* Technical Report Nr. 2332, TU Darmstadt, 2002.

[74] R. Sentis, *A Short Survey of Numerical Methods for Radiative Transfer Problems,* Journal de Physique, Colloque C7, Supplément nř 12, Tome 49,(1988), pp. 301–315.

[75] Shaukat Iqbal, *An Adaptive Finite Element Formulation of the Boltzmann-Type Neutron Transport Equation,* PhD Thesis, Faculty of Computer Science and En-

gineering, Ghulam Ishaq Khan Institute of Engineering Sciences and Technology, Pakistan, 2007.

[76] P. Sonneveld and M. B. van Gijzen, *IDR(s): a family of simple and fast algorithms for solving large nonsymmetric systems of linear equations*, Preprint, Delft Institute of Applied mathematics, Delft University of Technology, Delft, The Netherlands, March 2007.

[77] F. D. Swesty, D. C. Smolarski and P. E. Saylor. *A comparison of algorithms for the efficient solution of the linear systems arising from multigroup flux-limited diffusion problems*, The Astrophysical Journal Supplement Series, **182** (2004), pp. 369–387.

[78] Tamrabet A. and Kadem A. *A Combined Walsh Function and Sumudu Transform for Solving the Two-dimensional Neutron Transport Equation*, Int. Journal of Math. Analysis, **1** (2007), no. 9, pp. 409 - 421.

[79] A. Tizaoui, *Splitting Operator for Solving the Neutron Transport Equation in 1-D Spherical Geometry*, International Journal of Mathematics and Statistics, **1** (2007), no. A07, pp. 31-45.

[80] A. Tizaoui, *Polynomial Precondioning and the Generalized Minimal Residual Algorithm Solver for the 2-D Boltzmann Transport Equation*, C. R. Acad. Sci. Paris, Ser. I, **345** (2007), pp. 178-181.

[81] V.Trenoguine, *Analyse Fontionnelle*, Mir, Moscou, 1985.

[82] Vladimirov S. *Mathematical problems in the one-velocity theory of particle transport*, Technical report (translated from Transactions of the V.A. Steklov Mathematical Institute,61, 1961), AECL, Ontarion, 1963.

[83] N. J. Wager, *A Rapidly-Converging Alternative to Source Iteration for Solving the Discrete Ordinates Radiation Transport Equations in Slab Geometry*, Ph.D Dissertation, Air Force Institute of Technology, Air University, 2004.

[84] J. S. Warsa, M. Benzi, T. A. Wareing and J. E. Morel. *Fully consistent diffusion synthetic acceleration of linear discontinuous transport discretizations on three-dimensional unstructured meshes*, Nucl. Sci. Engr., **141** (2002), pp. 236–251.

[85] J. S. Warsa, M. Benzi, T. A. Wareing and J. E. Morel. *Preconditioning a mixed discontinuous finite element method for radiation diffusion*, Linear Algebra Appl., **11** (2004), pp. 795–811.

[86] G. Widmer, R. Hiptmair and C. Schwab. *Sparse Adaptive Finite Elements for Radiative Transfer*, Research Report No. 2007-01, Seminar für Angewandte Mathematik, Eidgenüssische Technische Hochschule, (2007).

[87] W. A. Wieselquist, *The Quasidiffusion Method for Transport Problems on Unstructured Meshes*: Ph.D Dissertation, Graduate Faculty, North Carolina State University, 2009.

[88] A. Yamamoto, Y. Kitamura, T. Ushio and N. Sugimura, *Convergence Improvement of Coarse Mesh Rebalance Method for Neutron Transport Calculations*, Journal of Nuclear Science and Technology, **41** (2004), No. 8, pp. 781–789.

[89] D. M. Young, *Iterative Solution of Large Linear Systems*, Academic Press, New York and London, 1971.